Ice Age Uncertainty

By Rolf A. F. Witzsche

Text copyright © Rolf A.F. Witzsche 2018

All rights reserved

Grand Solar Minimum
becomes the Ice Age

part 2:
Uncertainty

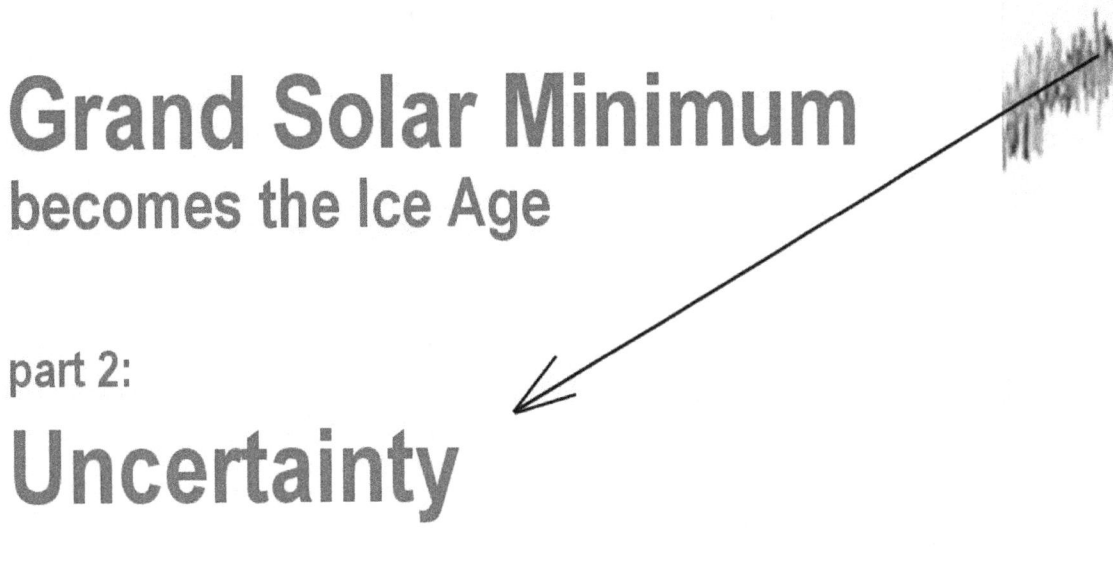

The book contains the transcript and images of the science exploration video by Rolf A. F. Witzsche, with the above title at: http://www.ice-age-ahead-iaa.ca-- The book is a part of the transcripts series.

Lead-in

The great Ice Age uncertainty is not located in the realm of Ice Age physics, or in our understanding of the principles involved, nor is it located in the lack of evidence. To the contrary, the principles are well understood, the physics is replicated in laboratory experiments, and the evidence that the Ice Age Challenge is real, is monumental. The great uncertainty lies in the court of society's response to what is known and understood and is already evident around the world. Society's response is so faint at the present time that it is almost non-existing. The question of whether there will be a response forthcoming that enables the whole of humanity to live past the onrushing crisis and create itself a bright future for it, for which the potential exists, remains yet to be answered.

As I have said, the physics are simple and the physics plainly apparent. We have observed many Grand Solar Minimum events in the past, and large Global Warming events for more than three thousand years. But we saw their amplitude and intervals shrinking. We saw the interglacial climate as a whole, diminishing, measured in ice cores. We also saw the level of solar activity measured in sunspot numbers and isotope ratios, diminishing at an ever-faster rate.

Now in more recent time, past the year-2000, we saw even the solar wind diminishing, measured at a rate of 30% per solar cycle, which is still ongoing. Even the 11-year solar cycle itself - the heart beat of the solar system - has slowed, from 11 to 13 years, and is increasing. The climate scene on Earth has thereby become a scene of wide-ranging anomalies, such as blizzards, floods, and droughts.

Ironically, what is deemed to be the next sequential Grand Solar Minimum event, anticipated for the 2030/40 timeframe, is not a cyclical event at all that the world will recover from. These events are history. The solar system no longer has the underlying support to recover from its ongoing collapse. While we will experience another Little Ice Age in the near future, the climate collapse towards it, that is already ongoing, has a different cause than the solar minimum events before. This different cause renders the now unfolding Solar Minimum event unrecoverable. This means that the event is not an event itself, but a transition phenomenon in the boundary zone to the start of the next Ice Age that renders the Earth largely uninhabitable for 90,000 years into the future.

It is worth repeating that the great uncertainty in all this is not located in the solar collapse process itself, and the corresponding collapse of the climate on Earth. These aspects are certain. The astrophysical processes are fairly predictable with scientific understanding of the principles involved. We have numerous measurements on hand that tell us that we have come to the end of the line of the interglacial world, with enormous consequences looming ahead for society. However, the challenge to master the consequences comes with grand opportunities in its wings for building us a New World with the technological power to avoid the consequences. While all of these factors are certain, our response to them is not. The response is presently a big ZERO. Still, the time remaining to us is our space for repentance, a time to choose life and develop a profound love for the wondrous humanity that we all share and to protect it with the technological power that we have already in the pocket.

We need to develop us a new humanist paradigm on this front that inspires us to built ourselves out of the new impending climate crisis, by building us a New World that the Ice Age Climate cannot touch. Only then, when we are committed to this path, will the uncertainty end.

Contents

The 'climate' cycles within our galaxy 11

The entire system is dynamically diminishing 12

The Interglacial Climate isn't merely diminishing 13

The last three of the big historic warming pulses 14

Diminishing 'landscape' of historic solar activity 15

Changing Grand Solar Minimum events 16

The dynamics of the progression 17

The multi-year ice-coring projects 18

Climate Change as the Sun gets weaker 19

The solar-system's clock is running slower 20

Mean temperature on the path to diminishing 21

The ever-changing Sun is a natural feature 22

Ice Ages started only in the last 2.7 million years 23

Ice Age glaciation was of shorter duration 24

Solar system powered for 15% of the time 25

For the remaining 85% of the time 26

Only the interglacial periods are suitable 27

An effort to explore why this is happening 28

Our living is presently totally dependent on the interglacial 29

We have large volumes of data collected 30

The time has come to explore 31

> Humanist-development paradigm 32

What has been accepted is archaic 33

Long history of resorting to mistaken premises 34

The hydrogen-gas Sun tale of dreams 35

The sunspots are dark 36

Tales of why a gas-Sun produces plasma winds	37
Hotter than the surface of the Sun itself	38
The Sun is a thousand times too light	39
The giant star UY Scuty	40
This star is totally impossible	41
The Sun that is NOT self-contradictory	42
A new humanist paradigm is needed	43
Science is a part of the humanist paradigm	44
➢ What causes the Interglacial Climate to diminish	45
A model that fits all the parameters, is the Plasma Sun	46
A Plasma Sun is a sphere of plasma	47
Plasma doesn't pile up on the Sun	48
The evidence is the sunlight that we see	49
The Plasma Sun is not its own master	50
A very-long 'solar' cycle of 100,000 years	51
Electric movement creates magnetic fields	52
Dynamic electromagnetic structure, the Primer Fields	53
When the Primer Fields are active	54
When the Sun hibernates	55
Hibernating for 85% of the time	56
Active with Primer Fields	57
The steep decline, near the off-point	58
Three types of different resonance effects	59
➢ The innermost resonance is slowing	60
The cycle time has become longer	61
Now this last bastion of stability is diminishing	62
The Primer Fields appear to have shifted	63

This increase happened dramatically fast 64

From 11 years to 13 years This is big! 65

Extreme uncertainties laying ahead 66

Deep climate changes are beginning to unfold 67

> The changing long cycles 68

Deep changes for 3,000 years already 69

Grand Solar Minimum 70

The 250-year cycles getting shorter 71

The 1,300-year intervals getting shorter 72

Caused by separate phenomena 73

Of two-fold nested Primer Fields 74

The Oort Cloud of nested asteroid fields 75

Nested plasma structures are possible 77

The toroidial shape 78

As giant plasma structures 79

The 250-year resonance cycle became faster 80

An orderly geometric regression 81

The 1,300-year cycle is now beating faster 82

> In the sub-visible domain 83

The cyclical Warming spike in the 1700s 84

No one was prepared at the time 85

All through the 1600s 86

Did the solar cycles stop? 87

No, the Sun didn't go to sleep during the Maunder Minimum 88

When the Sun is weak 89

Activity fluctuations occurred below the visible-threshold 90

A portion unfolds at the sub-visible level 91

They unfold at progressively lower levels 92

This is why the peak of cycle 24 is so low 93

As we look ahead into the near future 94

Grand solar Minimum in the 1970-80s 95

Minimal periods getting smaller 96

The Dalton Minimum of 1810 97

The expected Grand Solar Minimum of the 2030s and 40s 98

The long down-trend that we see here 99

The historic solar minimum cycles are so weak 100

The weakening of the Sun is symptomatic 101

The underlying level of the interstellar plasma density 102

At some point on the steep slope 103

We don't know what the minimal threshold level is 104

We can only say with a high degree of certainty 105

As it has been noted by Jaworowski 106

Get the fur coats out 107

➢ The 'sinking' of the 'Rock of Gibraltar' 108

The rate of the collapse of 30% in 10 years 109

The rate of diminishment of the solar-wind pressure 110

The solar-wind pressure cannot diminish below zero 111

A new phase in the solar collapse 112

The Sun's surface temperature on course to its 'sinking' 113

Uncertainty, is too mild a term 114

The rate of collapse in solar activity 115

Collapse measured in space by Ulysses 116

Solar wind zero level in the 2030s 117

Services the solar wind does provide 118

Solar wind to a kettle boiling off steam 119

The cooling of the Sun, will begin in the 2030s 120

The solar wind purging the plasma-fusion products 121

The fusion products will clog up the fusion cells 122

The solar wind aids in purging the cells 123

The cooling of the Sun will begin 124

Thanks to the moderating effect of the solar wind 125

In a self-escalating rate of diminishment 126

The self-escalating dynamic collapse 127

For how long the Sun can grow colder 128

The closer we come to the big Ice Age phase shift 129

➢ The song of the warning bells 130

The consequences of the weakening Sun 131

Sunlight includes a wide spectrum of light energy 132

Sunlight is absorbed in the atmosphere 133

In times when the sunspot numbers are high 134

The infrared portion of the sunlight 135

The water-vapor effect is the largest portion 136

When solar activity is low, ... 137

Ionized elements in the air 138

Increased cloudiness, in turn 139

The effect of cosmic rays on cloud nucleation 140

When cosmic rays were injected 141

Large climate fluctuations happen through the back door 142

➢ Not how we affect the climate 143

CO2 is comparable to a cat on the sidewalk 144

A 10-times lower absorption coefficient 145

CO_2 is 100 times less dense than water vapor	146
Greenhouse heat budget 45% of it as latent heat	147
Ominous 'writing on the wall'	148
➢ To turn 'bad' into 'gold'	149
To build us a New World	150
Life began in the sea	151
We can build secure agriculture afloat on the sea	152
Living in thousands of new cities	153
All nations working hand in hand	154
We cannot escape the coming Ice Planet Earth	155
Face the dimmer Sun with a song	156
We may not only walk on the water	157
When we raise ourselves up	158
The children of the world will be our joy and treasure	159
➢ More from the author:	160
14 Libraries of books and video productions	160

> ## Start of the exploration

The 'climate' cycles within our galaxy

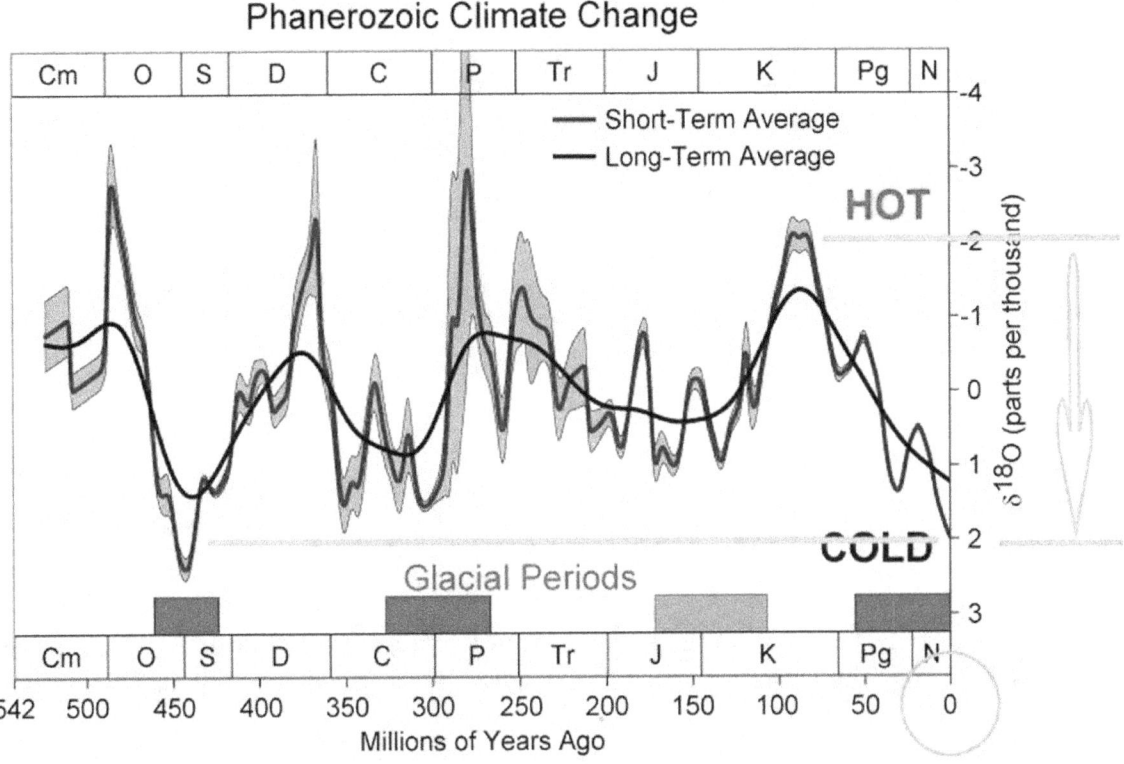

When we take a wide look at the climate history of the Earth, spanning the last 500 million years, a pattern of two long climate cycles, overlaid on each other, becomes recognizable, which reflect the 'climate' cycles within our galaxy.

It also becomes recognizable that the repetition rate of the individual cycles has remained constant throughout the entire long timeframe of over 500 million years. The strong constancy enables one to look forward in time and extrapolate quite accurately what the Earth's climate is and will be in the long run. Here we discover that the Earth's climate is presently at the weakest and coldest level in 440 million years. The weakness of the climate on Earth reflects the current weakness of the galactic system that affects our solar system.

The currently weak climate that we experience reflects the increasingly vulnerable dynamics of our solar-system and with it the weakness of our Sun.

Al this means that we live in precarious times, with a wide variety of uncertainties unfolding.

The entire system is dynamically diminishing

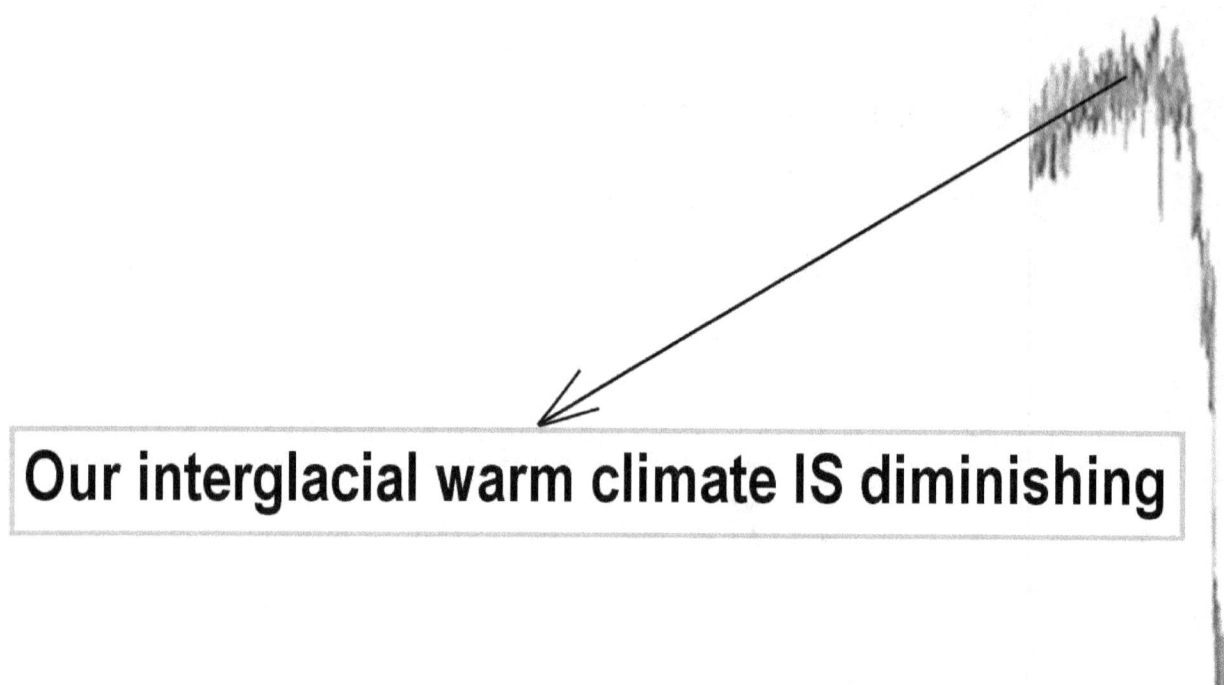

At the current state of the solar system, nothing at all is constant anymore. Every feature of the Sun and its supporting system is changing. The entire system is dynamically diminishing, and is diminishing rapidly.

The current climate on Earth, our interglacial climate, has itself been diminishing almost from the interglacial optimum onward, all the way to the present time.

The Interglacial Climate isn't merely diminishing

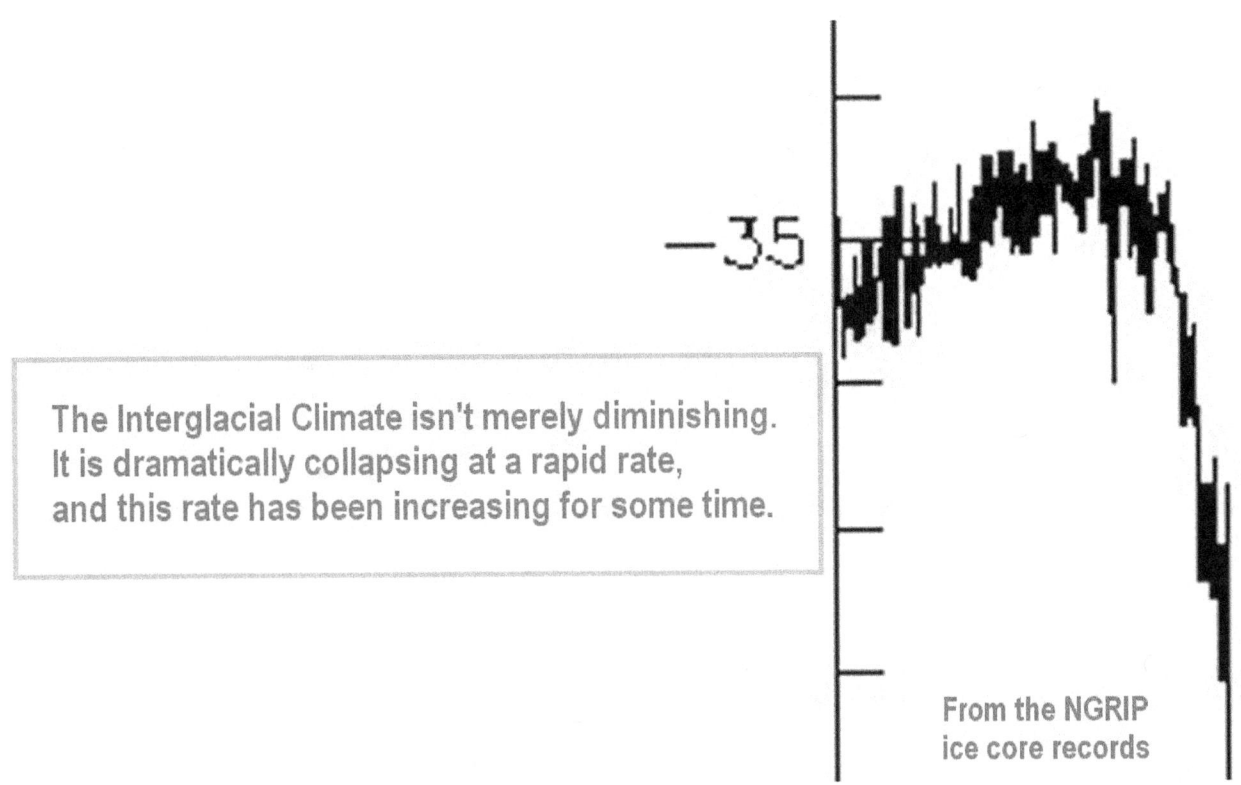

The Interglacial Climate isn't merely diminishing.
It is dramatically collapsing at a rapid rate,
and this rate has been increasing for some time.

From the NGRIP ice core records

And the Interglacial Climate isn't merely diminishing.

It is diminishing at a rapid rate, and this rate itself is increasing.

The last three of the big historic warming pulses

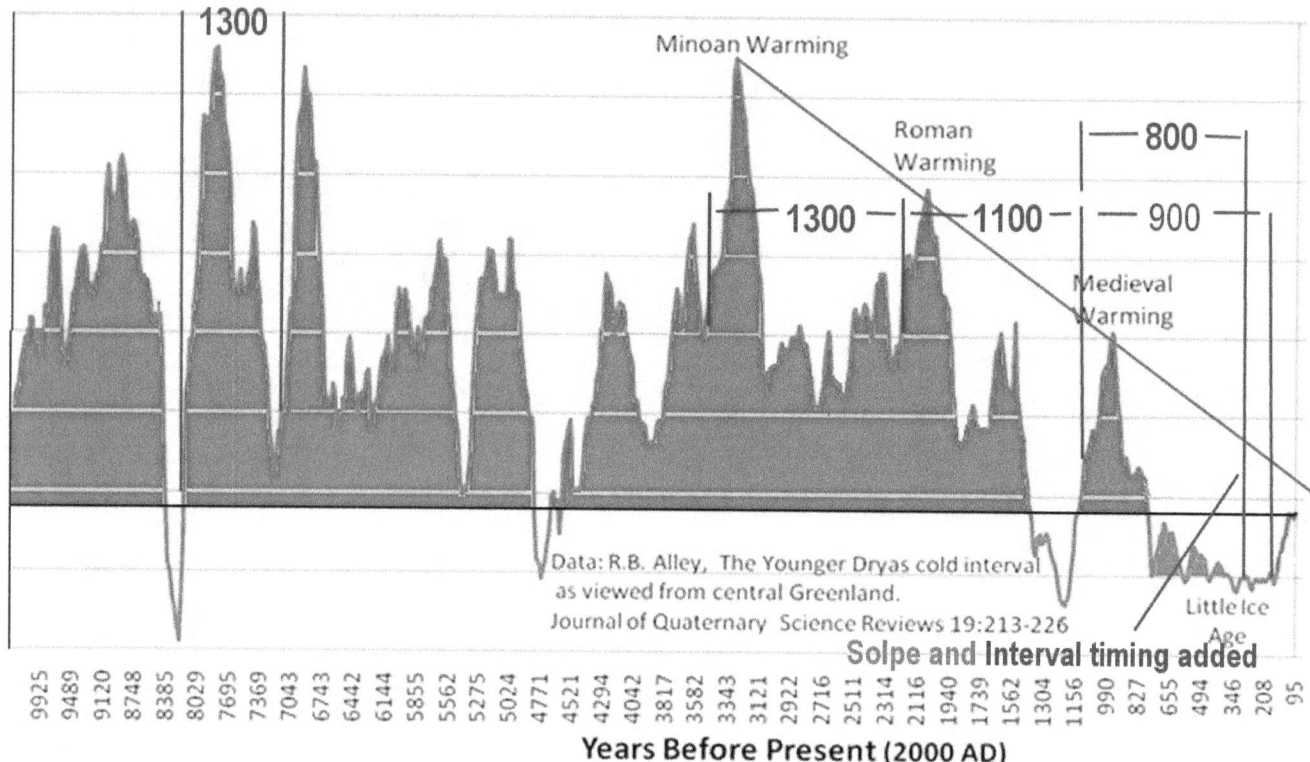

For example, the last three of the big historic warming pulses that have affected our civilization, have been getting progressively smaller in amplitude, and the intervals between the events have been getting shorter. Their intervals diminished from 1300 years, to 1100 years, and then to 800 years.

Diminishing 'landscape' of historic solar activity

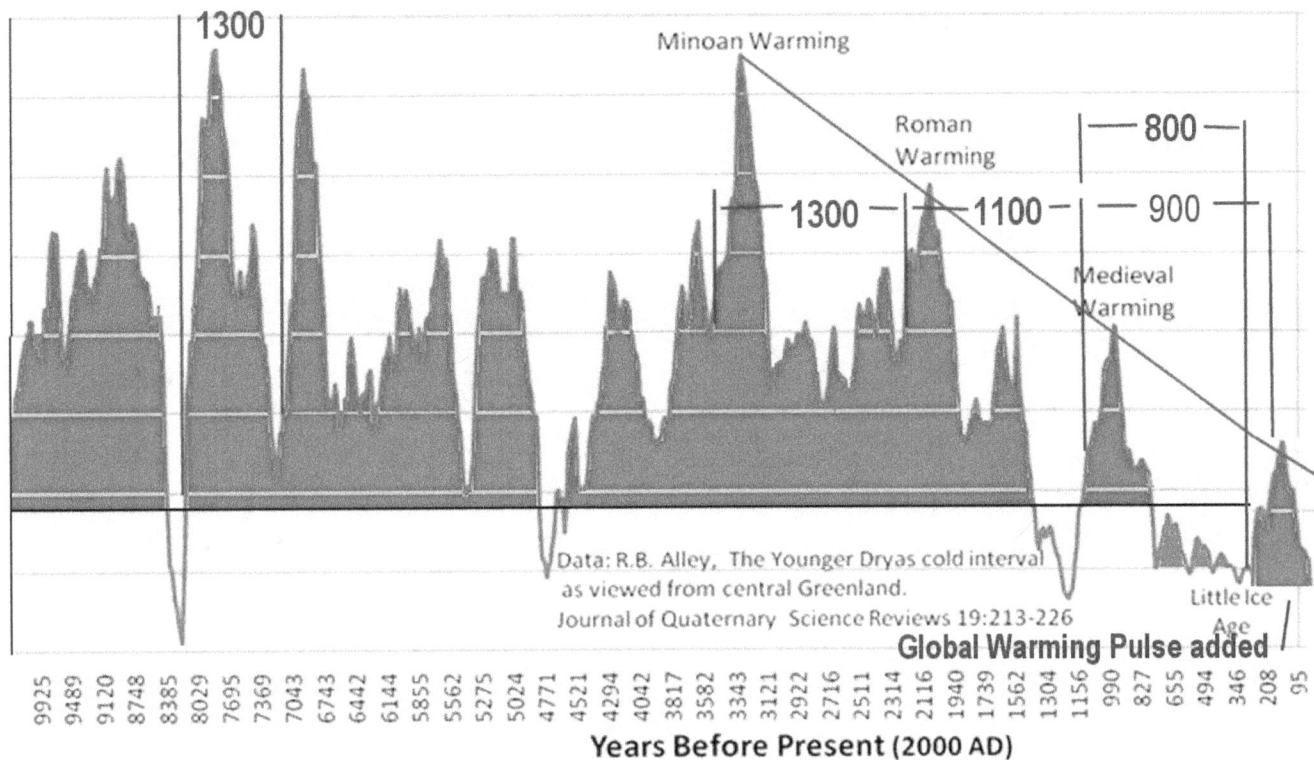

The now evermore diminishing 'landscape' of historic solar activity raises major uncertainties when we look forward in time.

Changing Grand Solar Minimum events

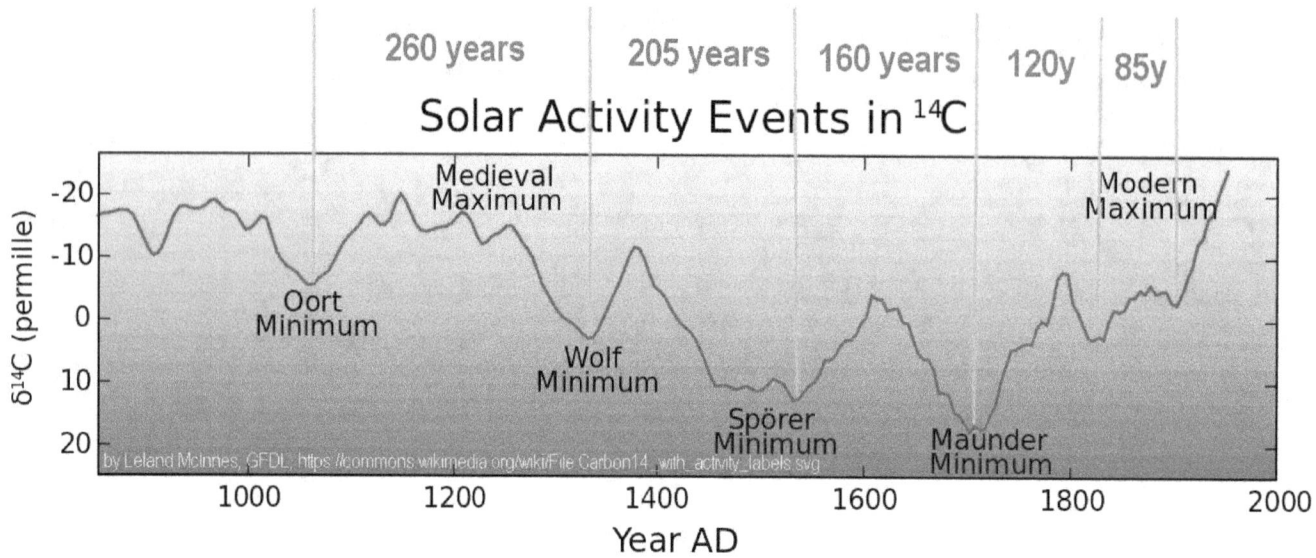

A similar case of shrinking intervals and amplitude is that of the changing so-called Grand Solar Minimum events. Both are changing so amazingly fast that nothing is confidently predictable anymore.

The dynamics of the progression

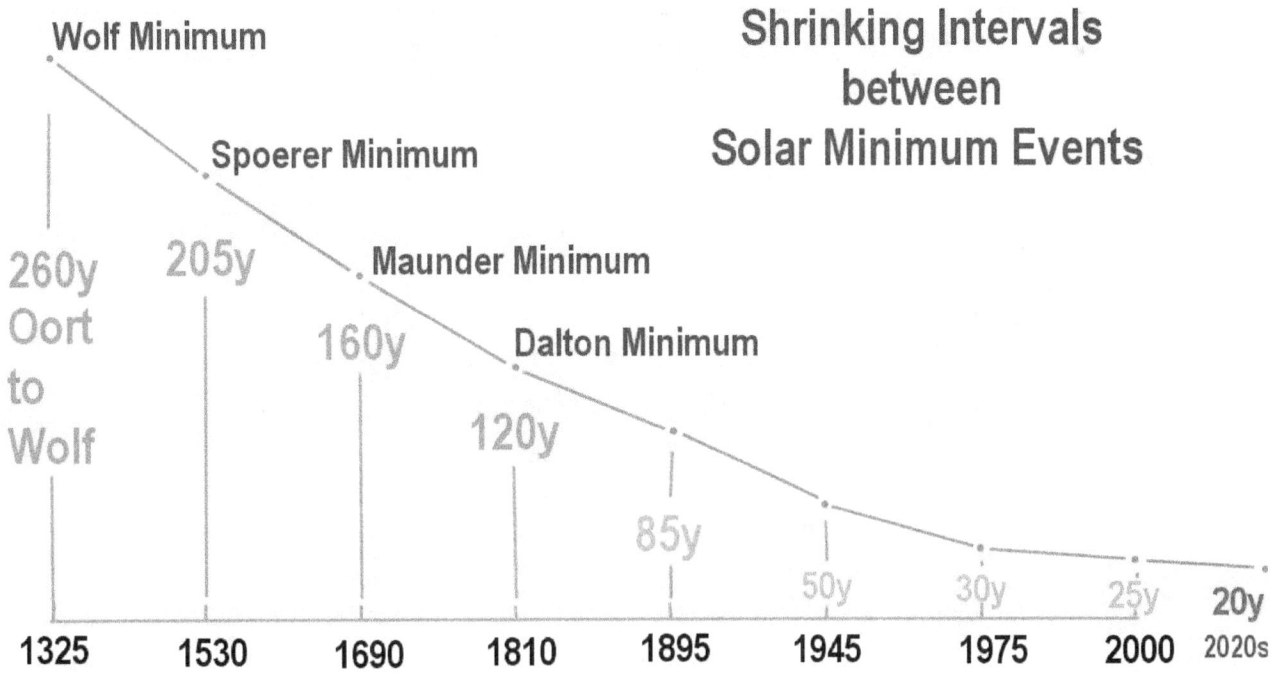

However, the dynamics of the progression give us a sense of the direction in which the Sun is moving. The intervals of the cyclical Grand Solar Minimum events are not merely diminishing in a random manner. The intervals between the events are getting shorter in a geometric progression. The geometric progression makes the outcome somewhat-more predictable.

The multi-year ice-coring projects

These recognitions are some of the benefits we derive from the multi-year ice-coring projects in Greenland and Antarctica.

It is our task as a human society to come to terms with what the discoveries mean for our living under the now rapidly changing Sun, and how to move with it, because those changes and their effects are hugely consequential.

Climate Change as the Sun gets weaker

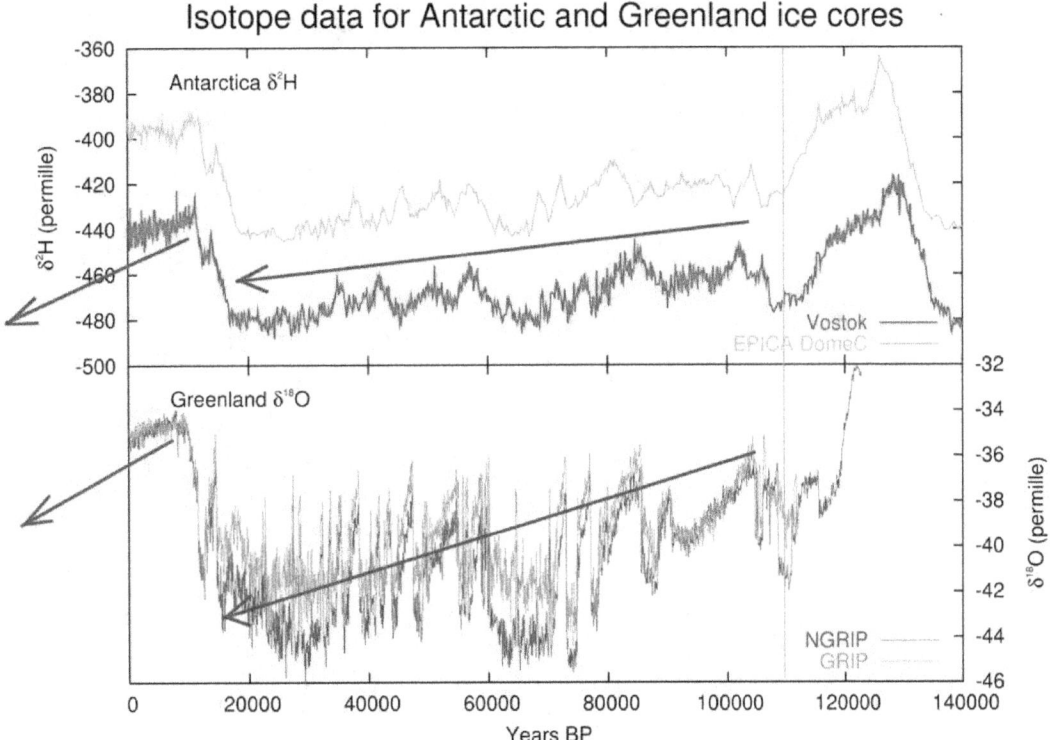

Climate Change is now evermore written in big letters as the Sun gets weaker.

The solar-system's clock is running slower

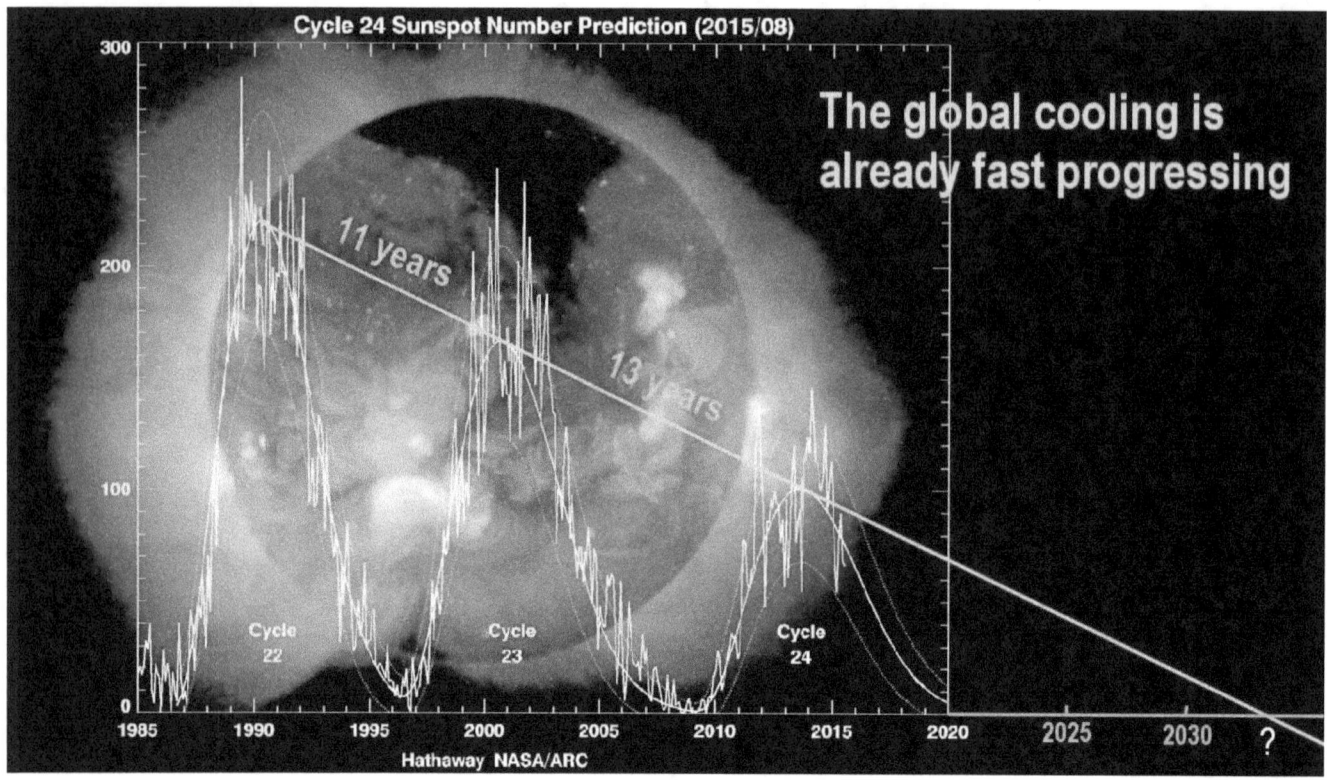

Even the very heart of the solar system, the 11-year solar cycle, which had been rock-solid in time, is now changing. The interval between the solar cycles is increasing. as the weakening is evermore felt. The solar-system's clock is running slower. The interval between the peak of cycle 23 and the peak of cycle 24, has increased from 11 years to 13 years with the weaker Sun. That's a big change.

Mean temperature on the path to diminishing

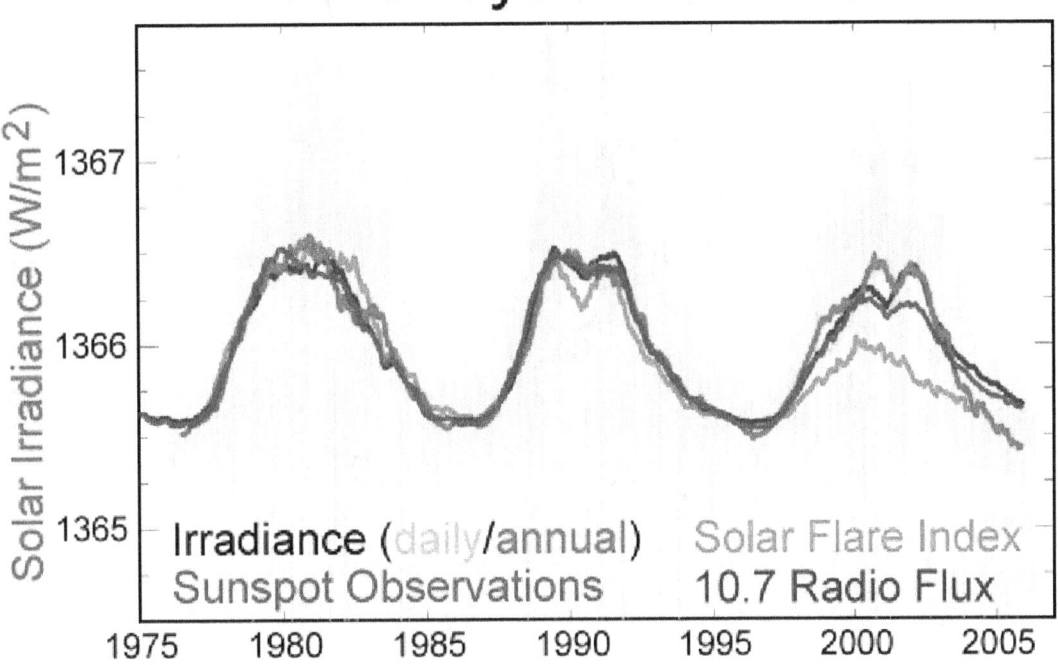

Even the mean temperature of the Sun, which hasn't changed since accurate measurements became possible, is on the path to diminishing potentially in the 2030s.

As the last shoe drops another inflection point is reached. But this shouldn't be cause for alarm.

The ever-changing Sun is a natural feature

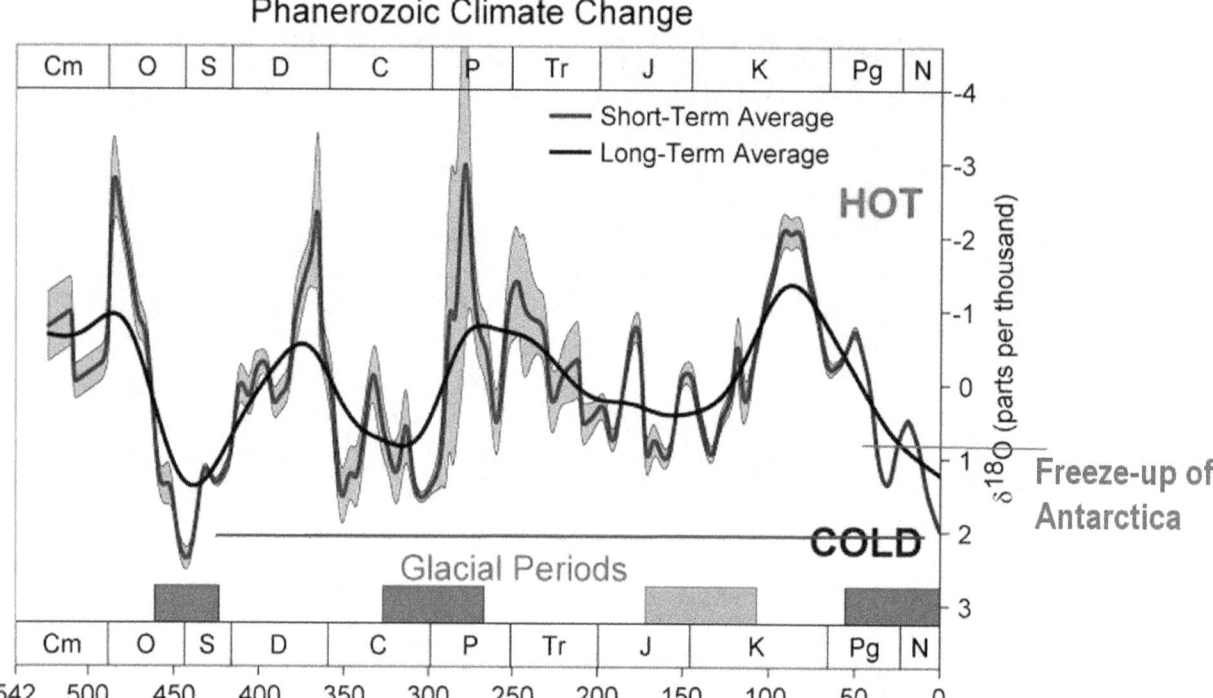

The ever-changing Sun is a natural feature of the nature of the universe, and of the dynamics of the solar system, especially now that the galactic dynamic system is presently at its weakest state in 440 million years.

Ice Ages started only in the last 2.7 million years

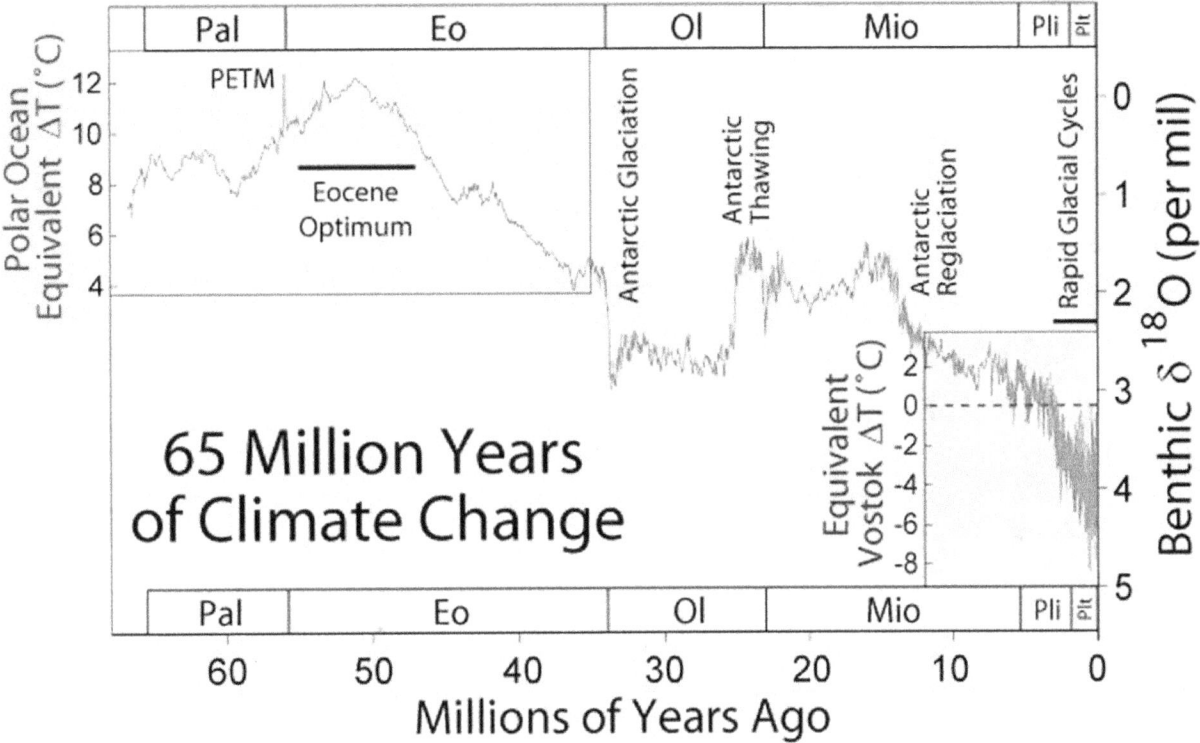

Two-thirds along the way from hot to cold, during the last 100 million years, Antarctica glaciated, then thawed out again, and glaciated once more around 12 million years ago. The Ice Ages started only in the last 2.7 million years of the still ongoing big galactic weakening.

Ice Age glaciation was of shorter duration

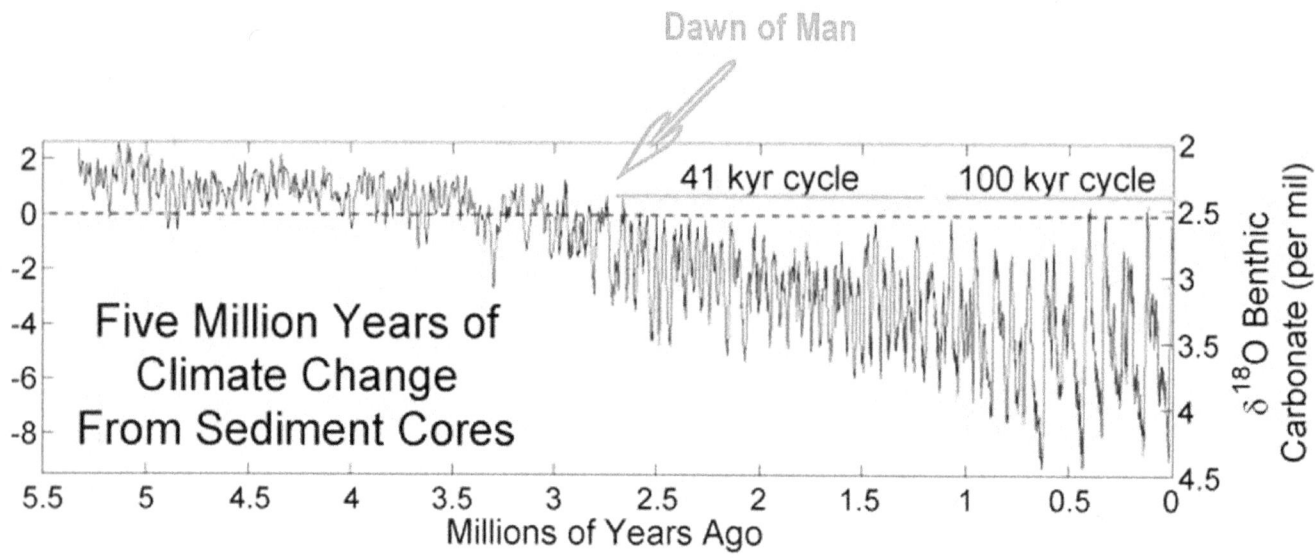

The Ice Age glaciation was of shorter duration at first in the range of 40,000 years, but as the galactic weakening increased, the glacial periods became longer and the interglacial periods shorter.

Solar system powered for 15% of the time

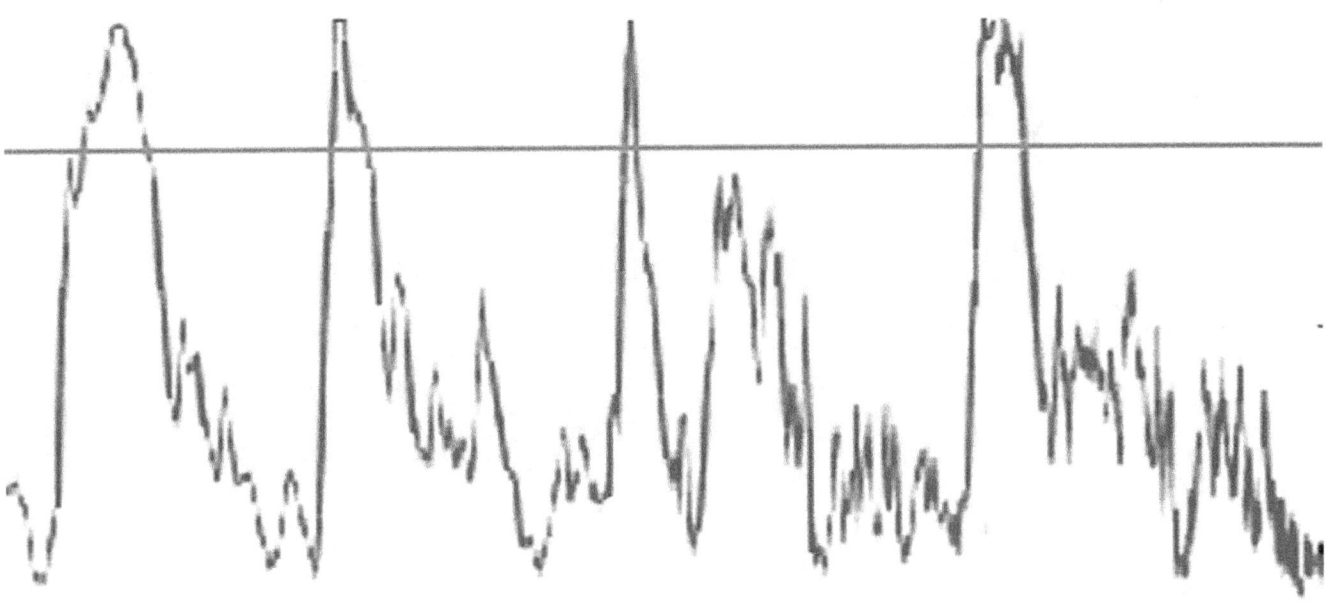

The galactic system has become so weak in the last half a million years that there is only enough strength left for the solar system to be fully powered for 15% of the time.

For the remaining 85% of the time

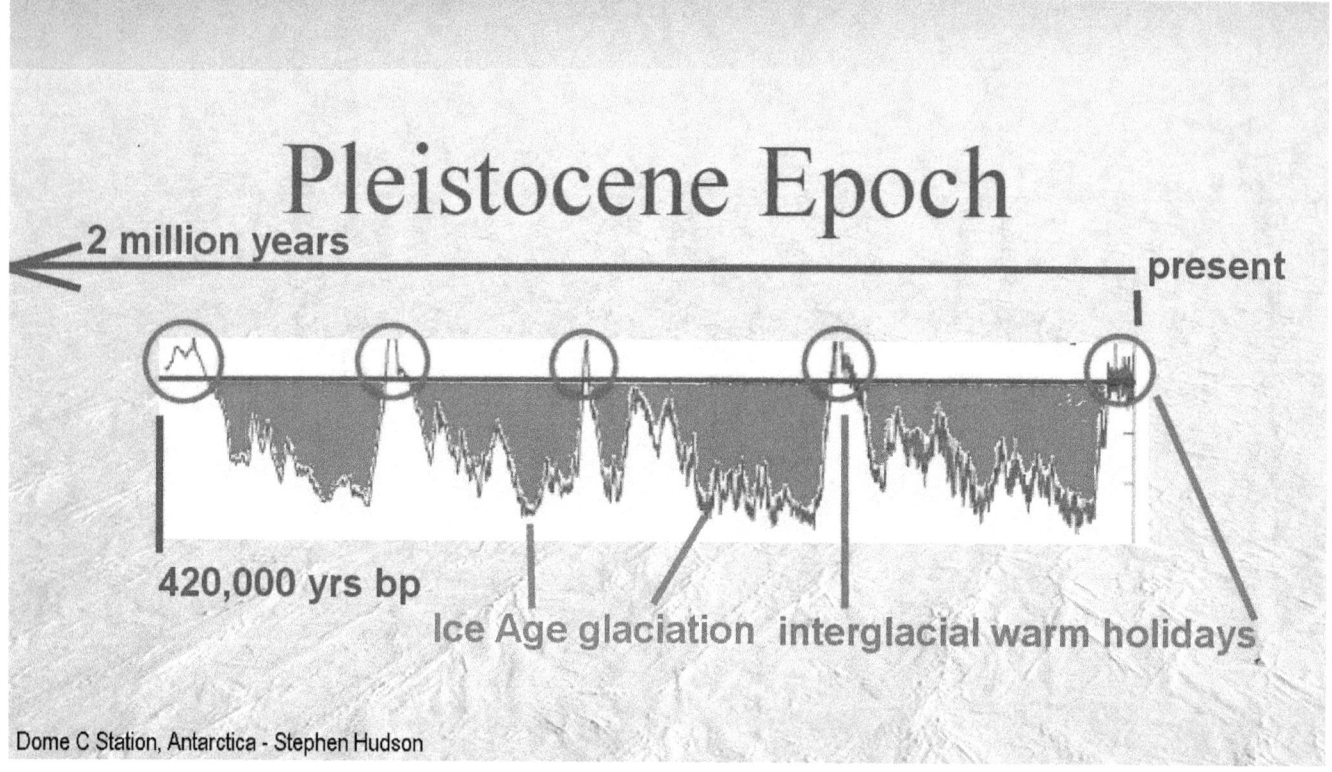

For the remaining 85% of the time, the ice core records tell us that the climate on Earth was an immensely cold and dry glaciation climate that renders the Earth largely uninhabitable.

Only the interglacial periods are suitable

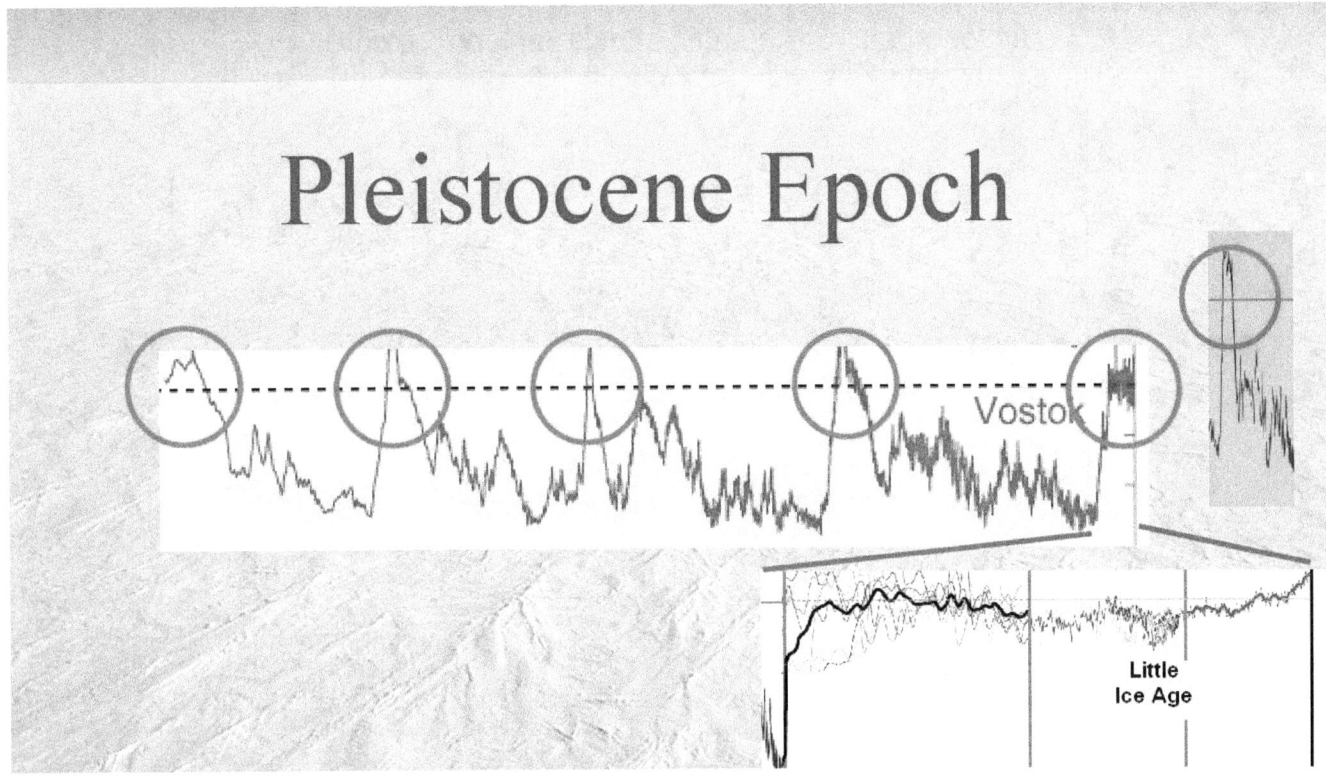

Only the interglacial periods are suitable for easy living as we have it today. The entire development of civilization occurred during the current interglacial period.

An effort to explore why this is happening

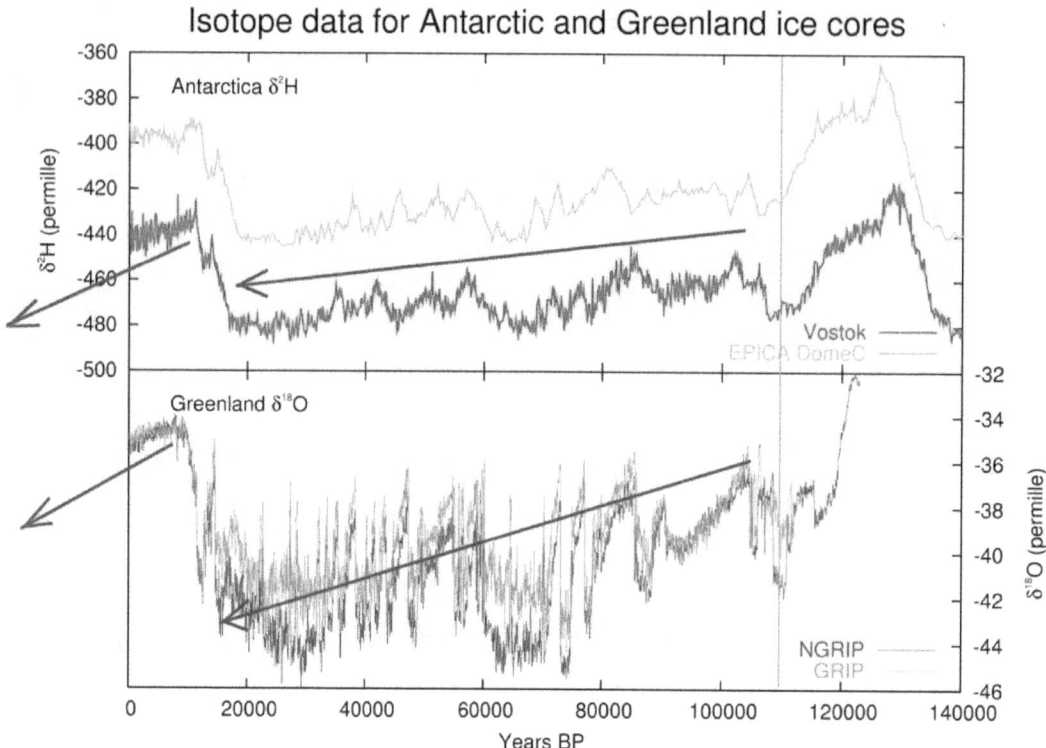

Since the modern ice core records tell us that the interglacial environment, in which we grew up, is rapidly weakening, it becomes imperative that we make an effort to explore why this is happening.

Our living is presently totally dependent on the interglacial

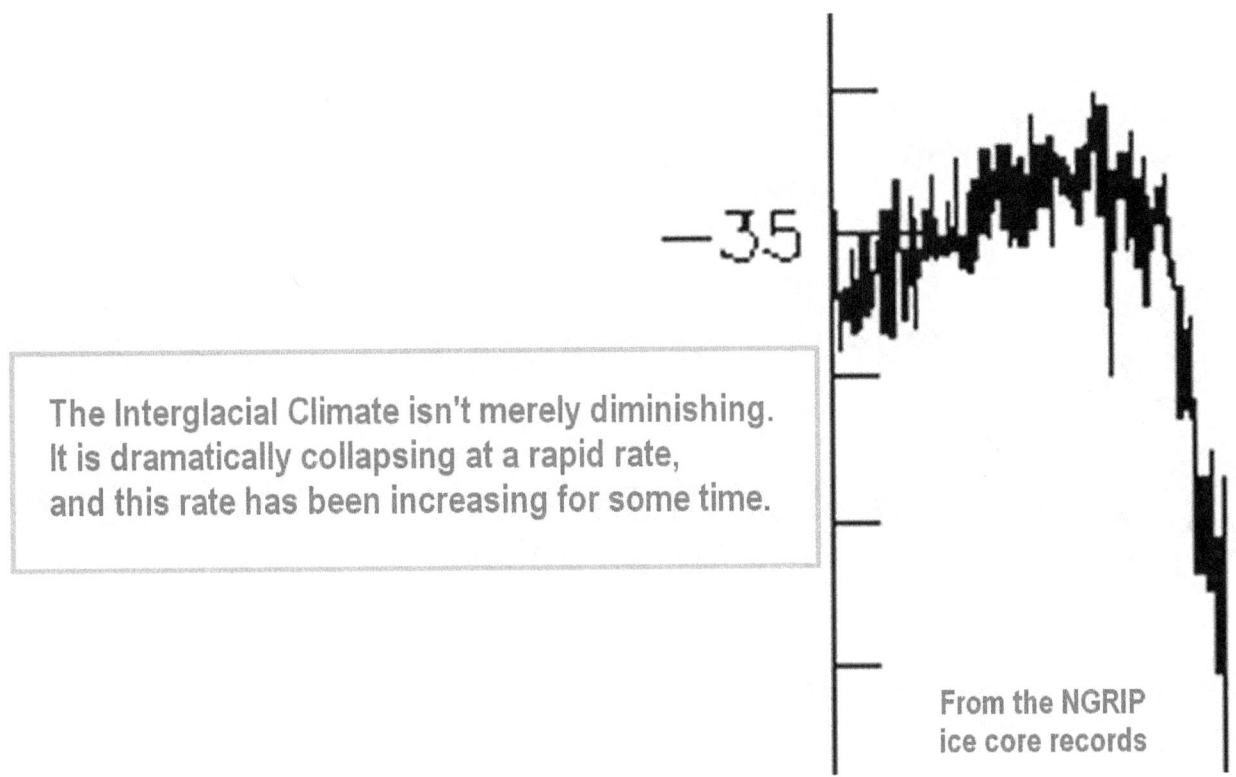

Since our living is presently totally dependent on the interglacial warm climate, it becomes especially imperative for us to explore why the interglacial climate is rapidly diminishing.

We have large volumes of data collected

We have large volumes of data collected about our climate history.

The time has come to explore

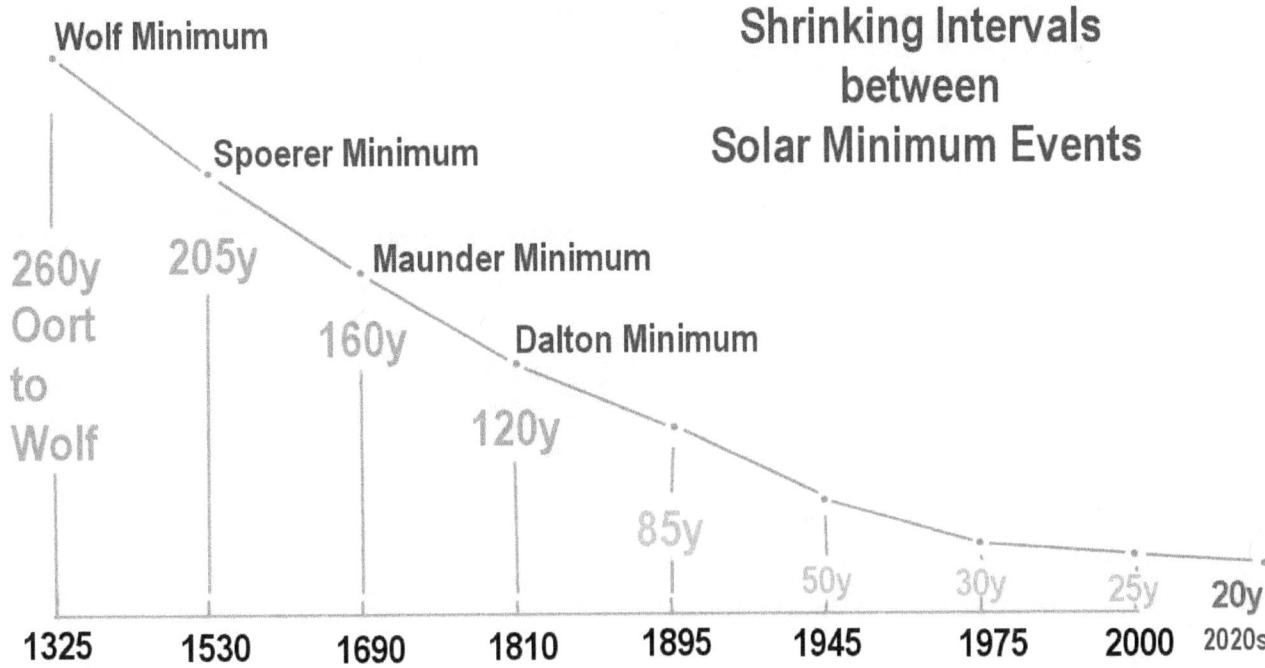

The time has come to explore what the collected data tells us about our future.

➤ **Humanist-development paradigm**

> The events-driven humanist-development paradigm

The events-driven humanist-development paradigm.

What has been accepted is archaic

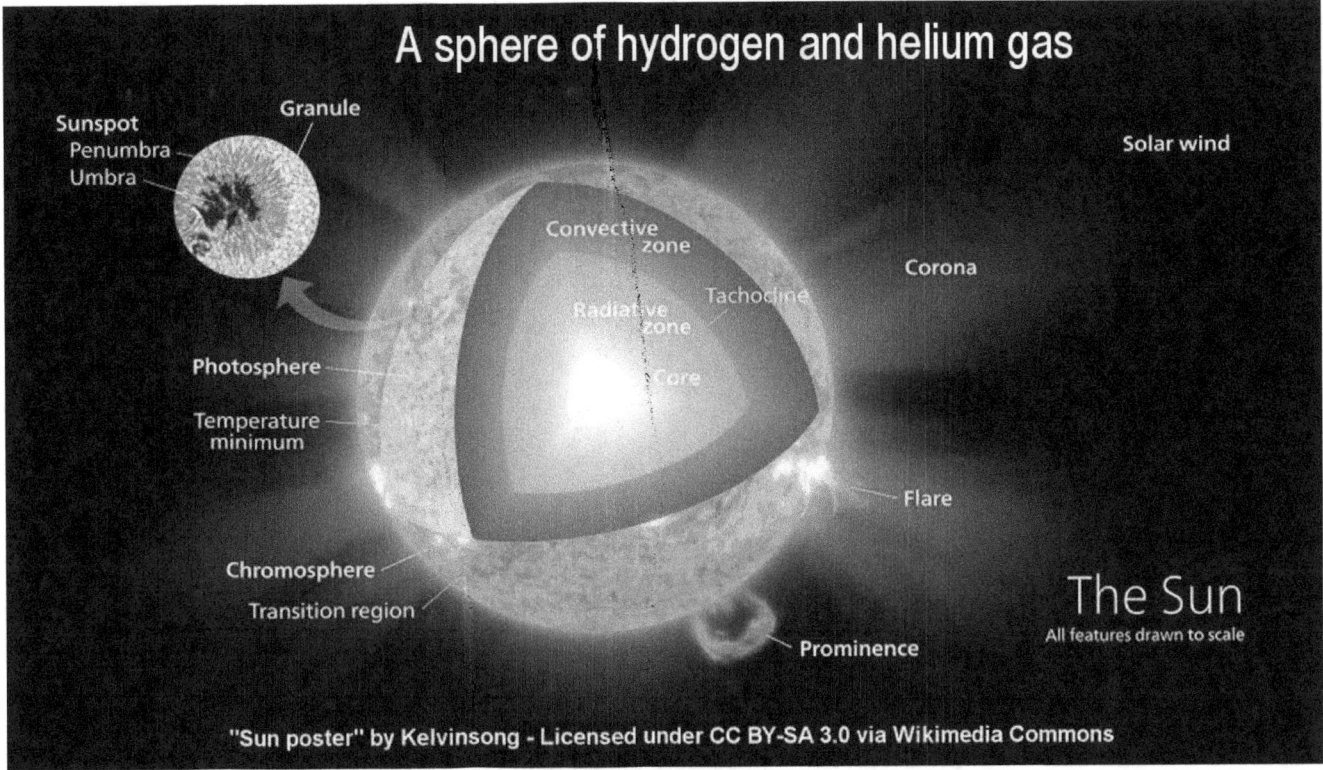

What we see happening in the numerous records of actual physical historic events, is ironically not possible to perceive on the widely accepted platform of the Sun that has become mainstream science.

What has been accepted is archaic. None of the large changes that have been discovered are compatible with the mainstream model for an invariable Sun.

Even in itself, the theorized model is so full of holes that it is self-evidently false. The large dynamics events that have been discovered about the solar system and the Sun, demand a dramatic advance in the science of solar physics, and an advance in the humanist paradigm that enables us to respond to what has already been discovered. Here, advanced and truthful science aids us, while archaic science that may be intentionally false, blocks our responses.

As I said, none of the discovered and fast diminishing climate phenomena are possible on the platform of the internally powered hydrogen-sun theory, which, according to the theory itself cannot change, much less change rapidly.

It is theorized under the archaic theory that the heat from the core of the Sun takes 30 million years to come to the surface by convection, and roughly 15000 years by photon emissions, absorption and scattering. Such a Sun cannot change as rapidly and extensively as we have observed the real Sun to change.

Long history of resorting to mistaken premises

But then, astronomy has had a long history of resorting to mistaken premises in order to meet politically imposed objectives, such as those that Ptolemy's epicycles were invented for.

The hydrogen-gas Sun tale of dreams

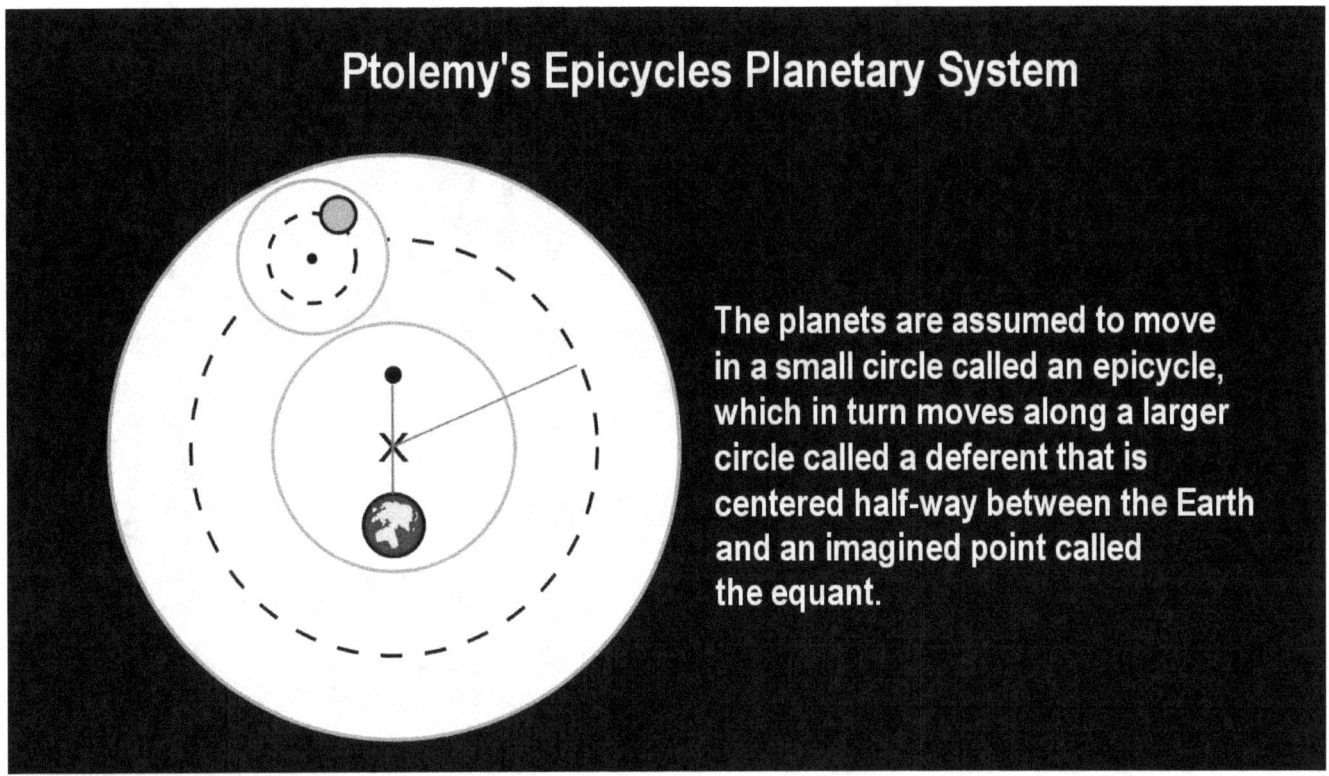

The hydrogen-gas Sun theory is a imaginary tale of dreams of a similar inventive type as Ptolemy's epicycles were. The time has come for a new humanist paradigm of scientific development to be developed that actually relies on verifiable physics instead of on mysticism.

The sunspots are dark

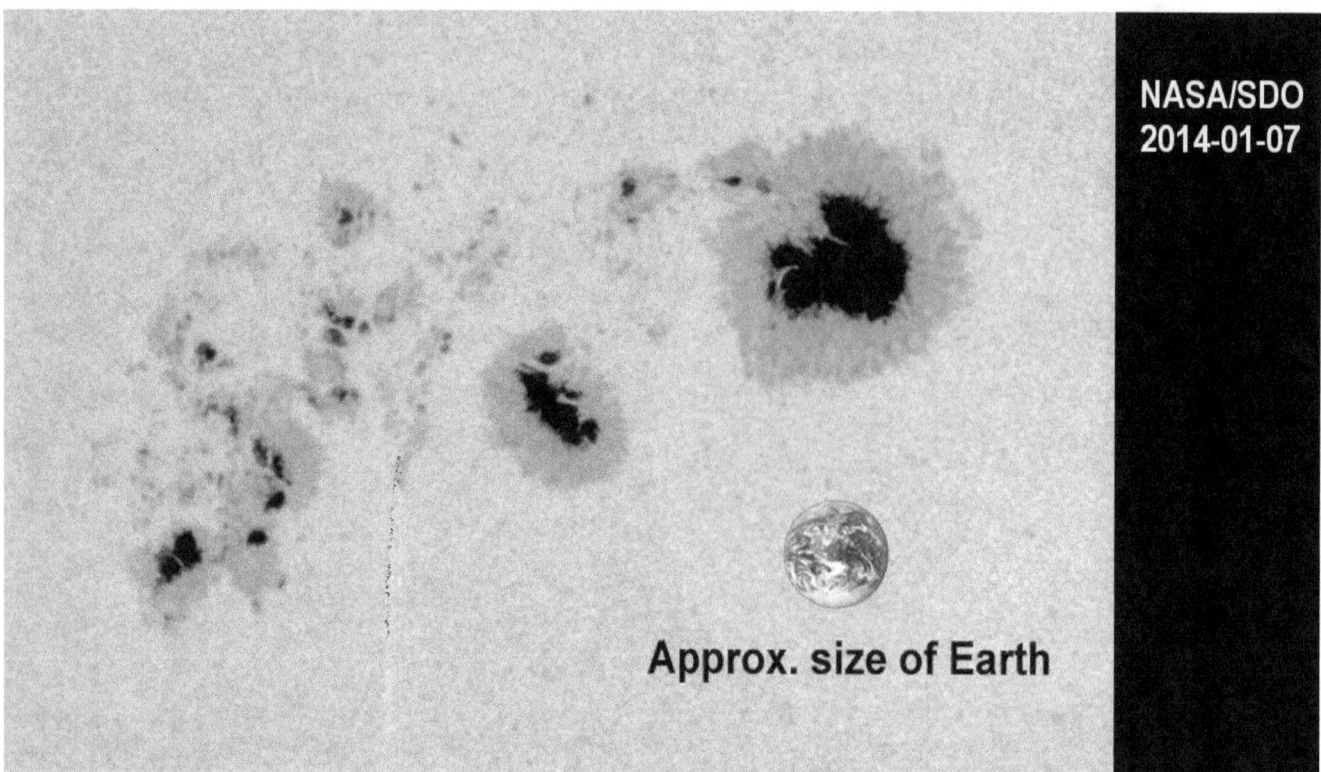

When we look at the Sun with a telescope, we see rather plainly that the sunspots are dark, while the opposite should be the case if the Sun was heated from the inside. We are told that this is so because mythical dark energy that blocks light, streams through the sunspots. And this is only a part of the theorized mysticism that is spun around the Sun, that even children would find hard to belief.

Tales of why a gas-Sun produces plasma winds

http://www.zam.fme.vutbr.cz/~druck/Eclipse/ - an example of the amazing solar eclipse photography of Milloslav Druckmueller

The mystery of wave actions is another fairy tales of why a gas-Sun produces plasma winds that mysteriously stream away from the Sun against the force of its gravity and accelerate as they go, which shouldn't be possible under the archaic theory.

Hotter than the surface of the Sun itself

Solar corona — by Luc Viatour / www.Lucnix.be - wikipedia

Similarly the mystery of wave actions bids us to imagine that the corona around the Sun can be many hundredths of times hotter than the surface of the Sun itself.

The Sun is a thousand times too light

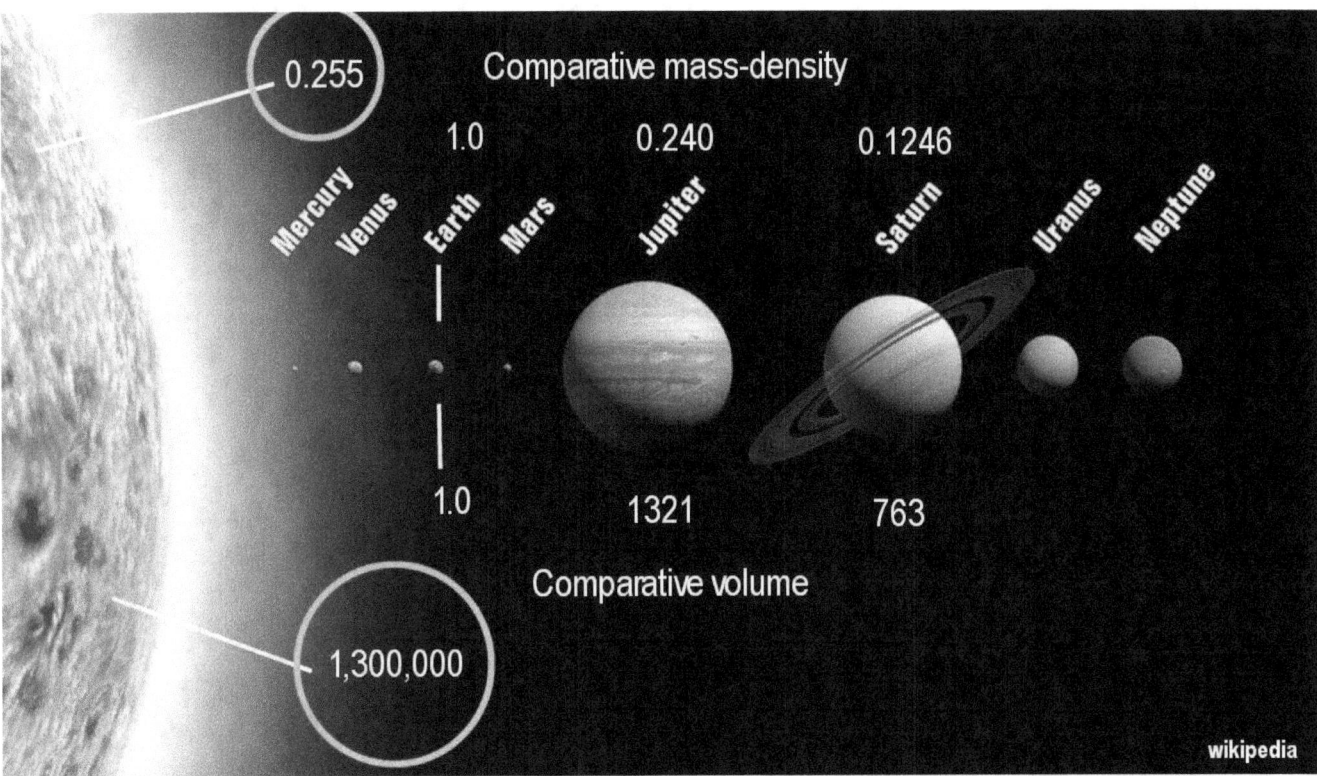

Furthermore the mystery of electron degeneration bids us to belief that a gas sphere of the size of the Sun is actually able to exist with a theorized gas-density at its core that is 150-times denser than water, while in comparison with other gas planets the Sun is a thousand times too light, and more than that are the giant stars.

The giant star UY Scuty

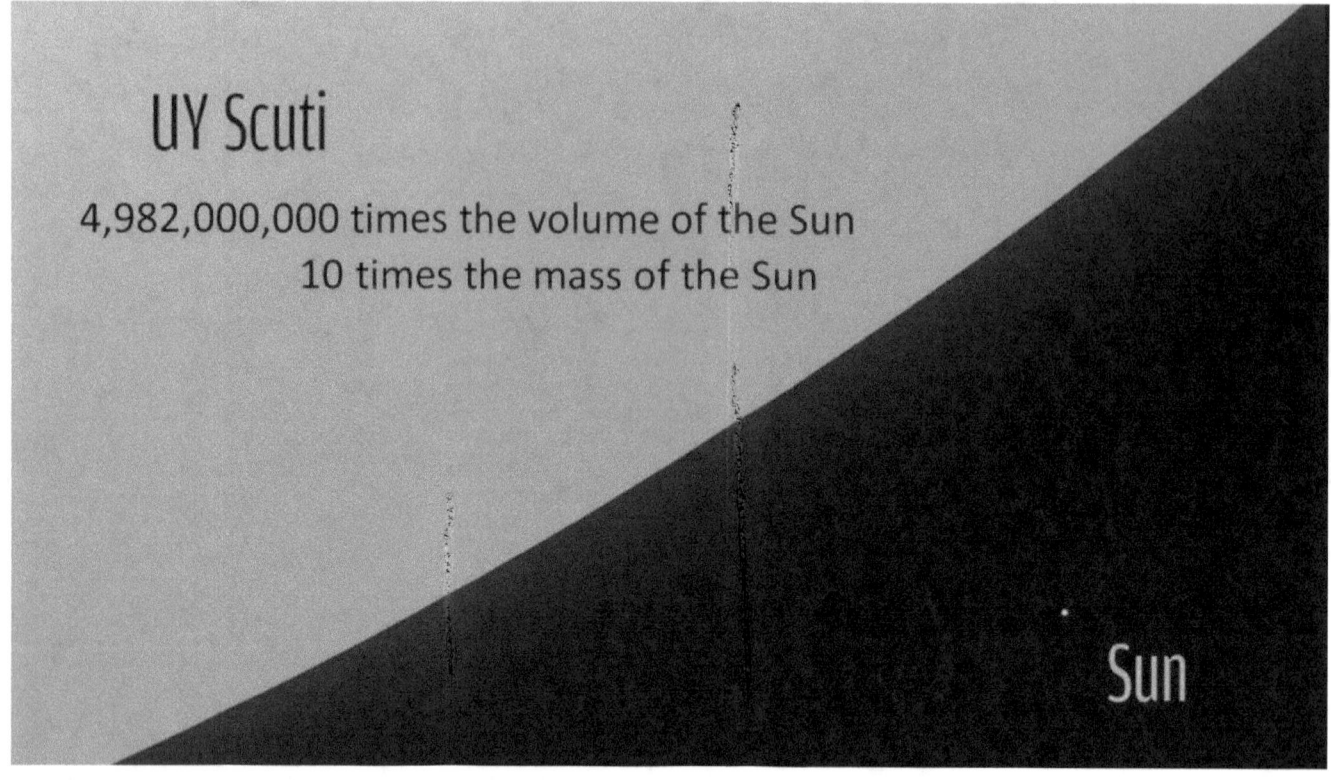

The giant star UY Scuty is said to contain 10-times the mass of our Sun. However, this mass is spread out across the star's 5 billion-times larger volume. A researcher commented that this mass is spread so thin that the star is practically a vacuum.

This star is totally impossible

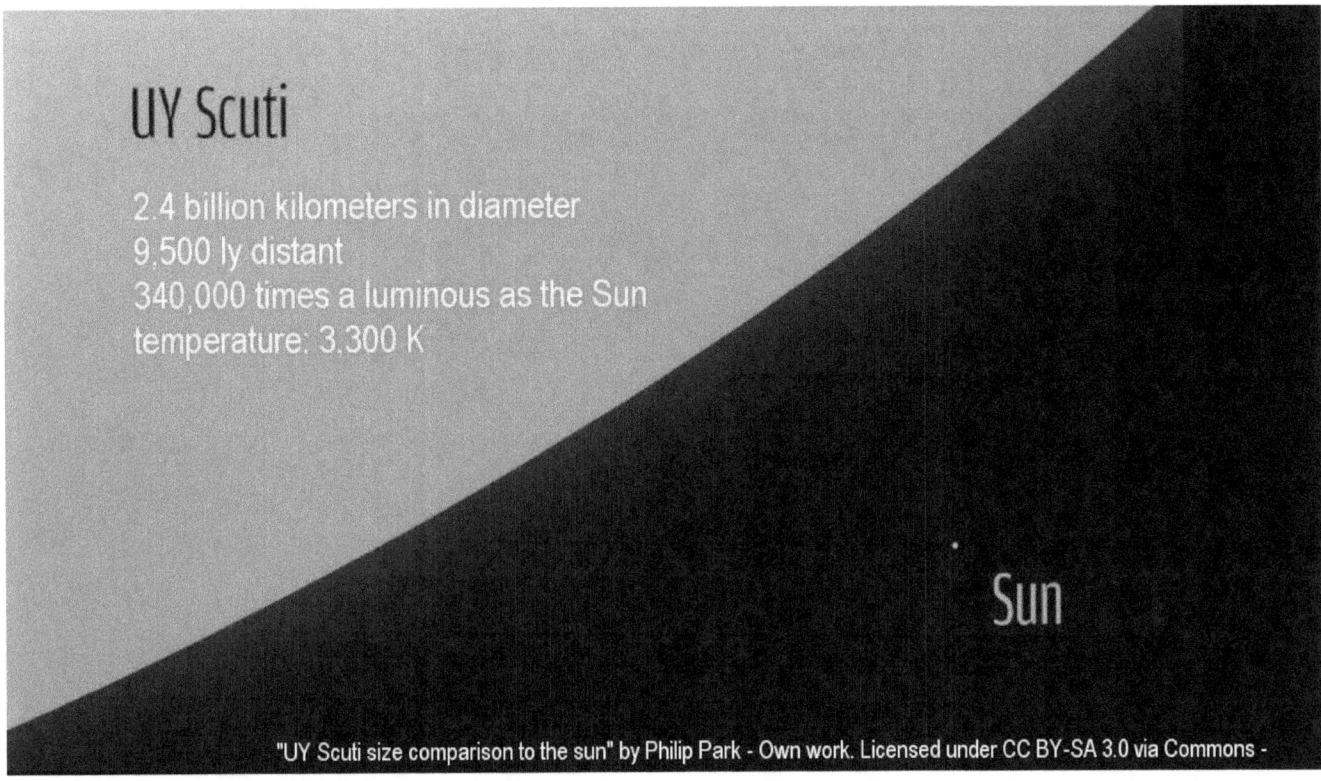

Yet physical measurements indicate that this vacuum-star outshines our Sun 340,000-fold.

Every aspect of this star is totally impossible under the gas-compression solar model, no matter what mysteries one attaches to this model. Nevertheless the star does exist and operates amazingly well. This means that the theorized model is self-evidently false.

The Sun that is NOT self-contradictory

The time has come to let go of the mysticism and the evidently false models built on them, and to develop a scientific model for the Sun that is NOT self-contradictory, and is in addition able to support all the amazing physical discoveries about the Sun that have been made in recent years.

A new humanist paradigm is needed

A new humanist paradigm is needed that focuses on truthfulness in science; and on the discovery of universal physical principles that enable us to intelligently respond to the solar events and their consequences that shape our world.

Science is a part of the humanist paradigm

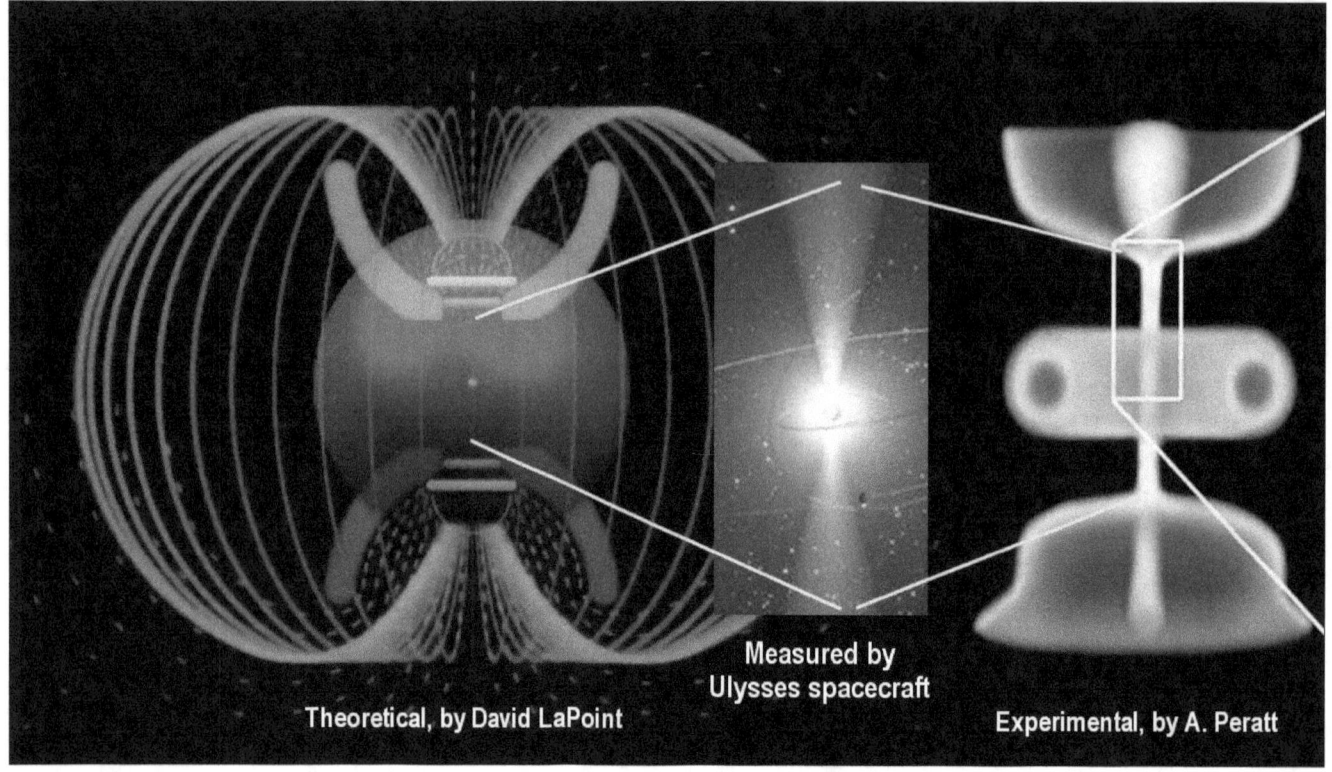

Science is a part of the humanist paradigm. It has not only created civilization, but lets us see the unseen and respond without doubt and fear to what we have discovered about the Sun, which we need to develop into a model that can aid us to respond in our living to what is real.

➢ **What causes the Interglacial Climate to diminish**

> Let's begin by exploring what causes the Interglacial Climate to diminish

Let's begin by exploring. What causes the Interglacial Climate to diminish?

A model that fits all the parameters, is the Plasma Sun

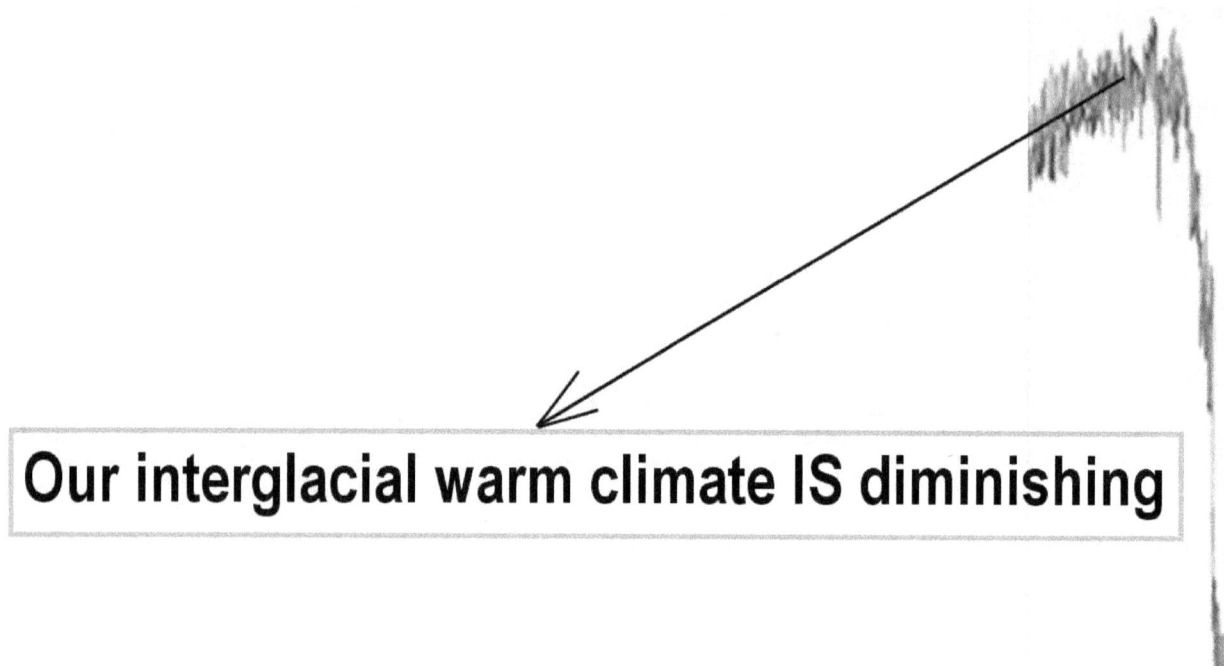

One is challenged to discover a model that can produce all the numerous diverse dynamics in physical events that we have discovered.

The only platform that offers such a model that fits all the parameters, is the Plasma Sun platform.

A Plasma Sun is a sphere of plasma

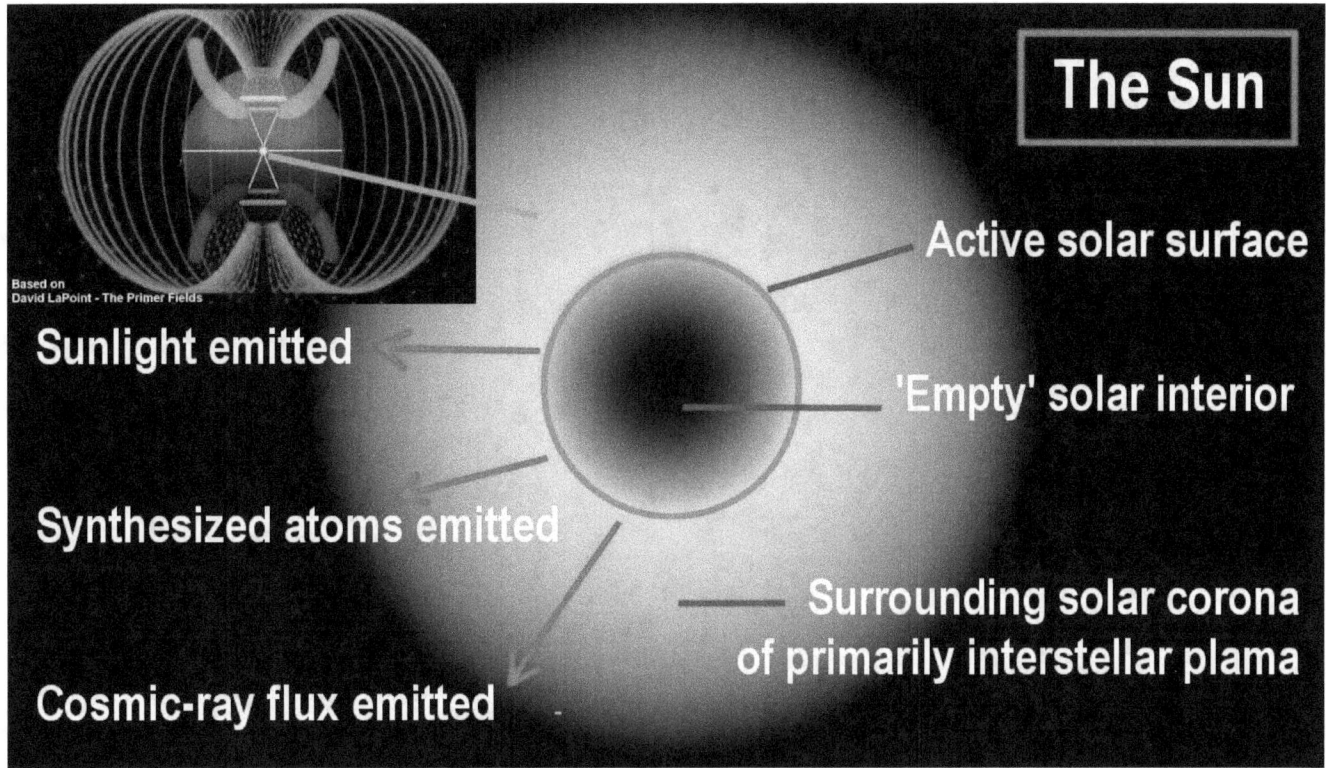

A Plasma Sun is a sphere of plasma. Plasma is the name for electrons and protons existing in unbound form, primarily in space. When large quantities are brought together into a sphere the size of the Sun, the lighter electrons that swarm around the protons migrate to the surface, away from the center of gravity. The change of distribution enables the protons at the center to repel each more, which gives the center a low mass density, and the surface an extremely high-mass density and high electron-density. Such a Sun presents to the cosmos an extremely high electric potential, which attracts plasma from interstellar space.

Plasma doesn't pile up on the Sun

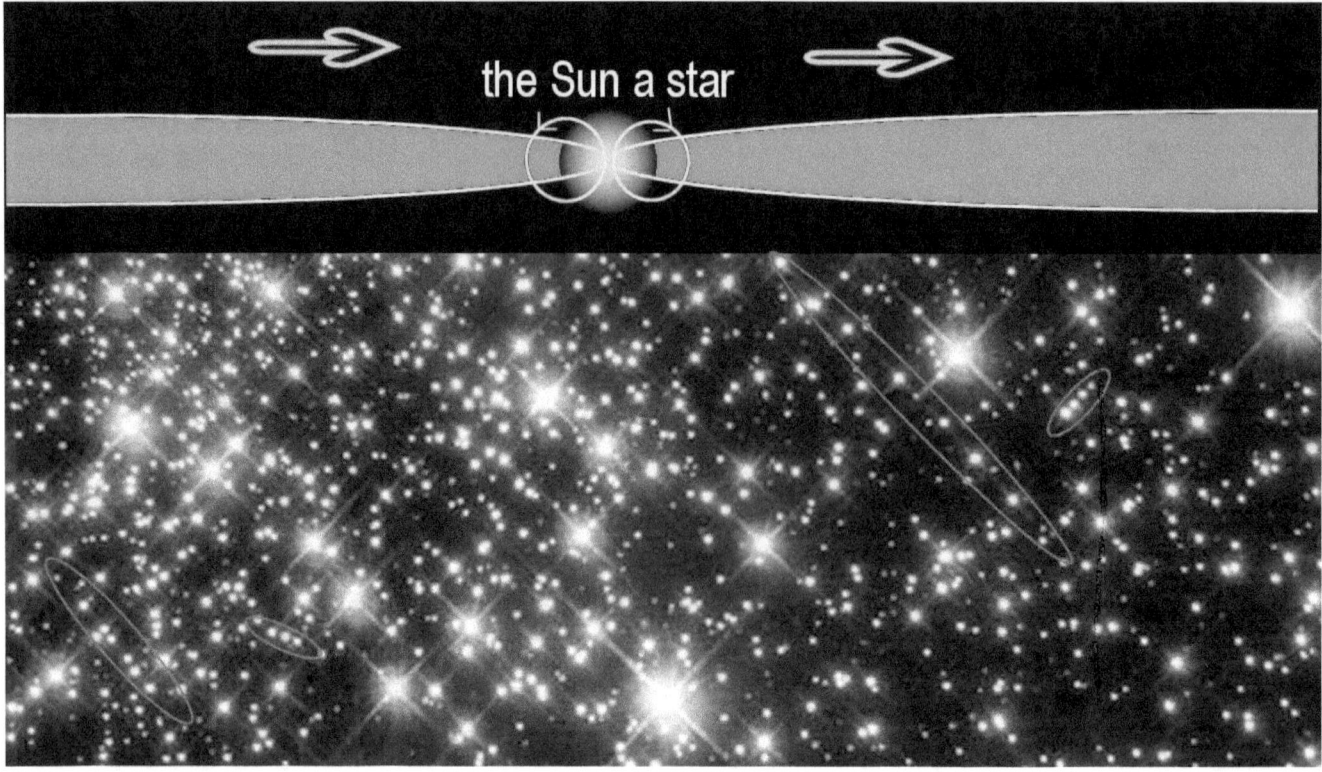

The interstellar plasma flows onto the Sun in long plasma streams, but the plasma doesn't pile up on the Sun. The Sun consumes a portion of it in an electrodynamics process of atomic synthesis that binds the plasma up into electrically neutral atomic structures. The plasma that the Sun doesn't consume, simply flows on.

The evidence is the sunlight that we see

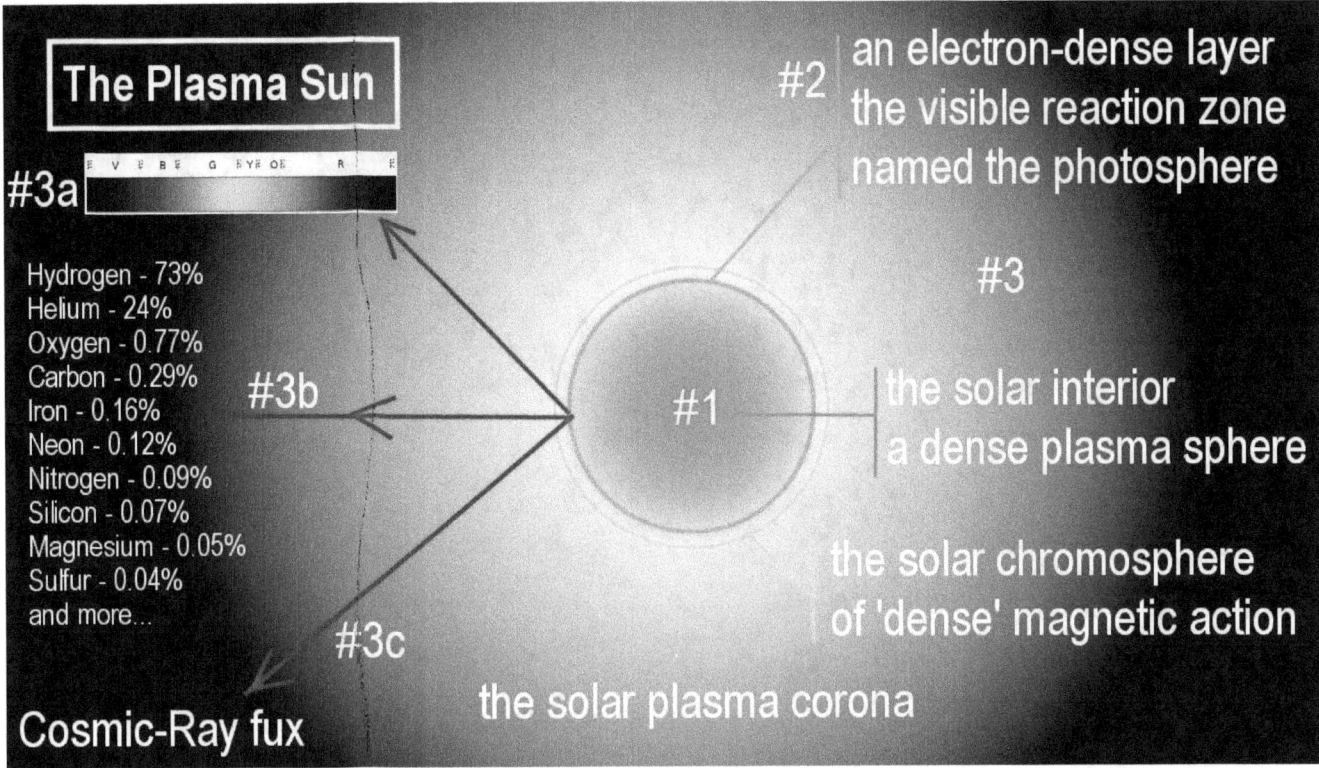

Intense action happens on the surface of the Sun. All atoms that the planets are made of, were created in this process on the surface of the Sun, by which interstellar plasma is massively consumed. The process is highly energetic, so that the created atoms are intensely energized by it and light up the Sun. All atoms emit energy in the form of photons of light, when they become energized. The evidence that this happens on the Sun is the sunlight that we see.

The Plasma Sun is not its own master

The nature of the Plasma Sun is such that it is not its own master. It responds to what flows into it. When the inflow is strong, the Sun becomes extremely active. When it is week, the Sun becomes 'quiet'.

A very-long 'solar' cycle of 100,000 years

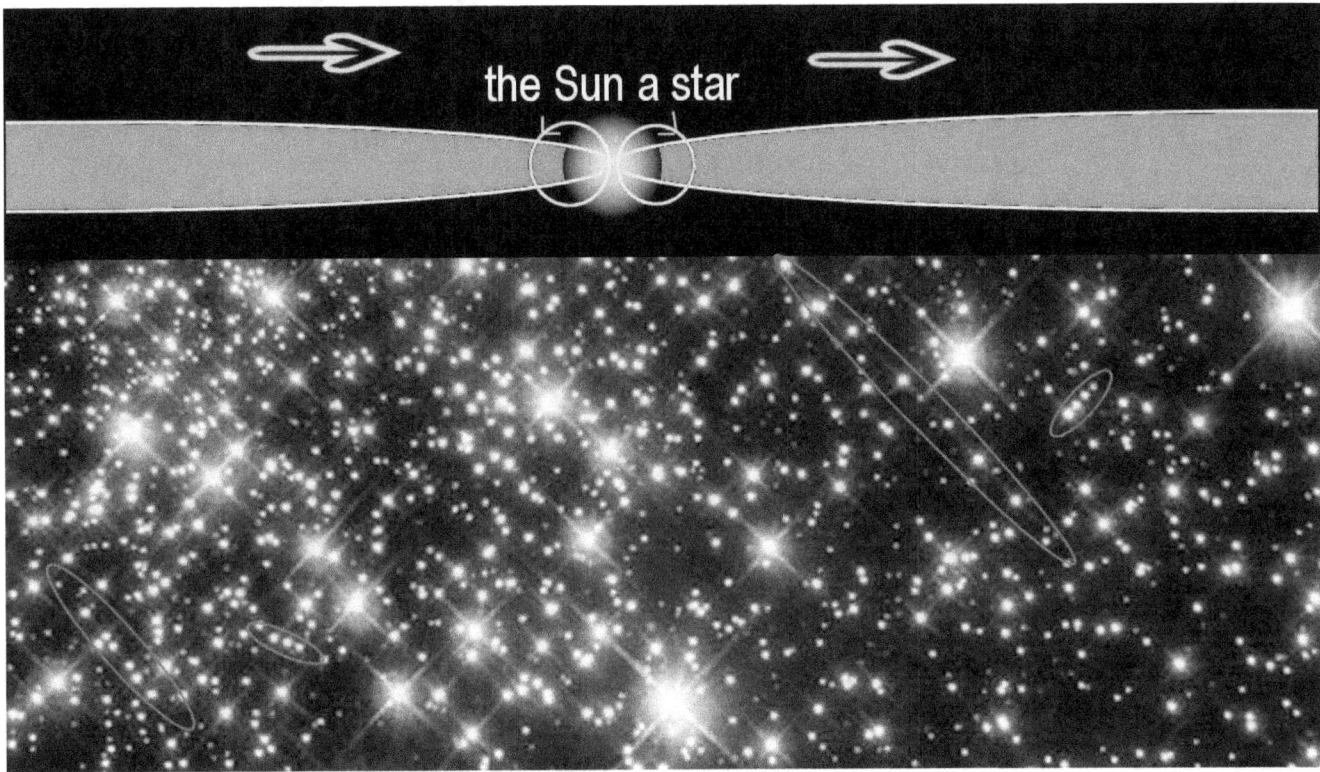

I have presented in previous videos, that our Sun and its solar system is located at a node point on an interstellar plasma stream.

The extremely long dimension of interstellar plasma streams generates a electric resonance in the plasma stream with a very-long 'solar' cycle, in the order of 100,000 years. All plasma structures have a specific resonance according to their size.

Electric movement creates magnetic fields

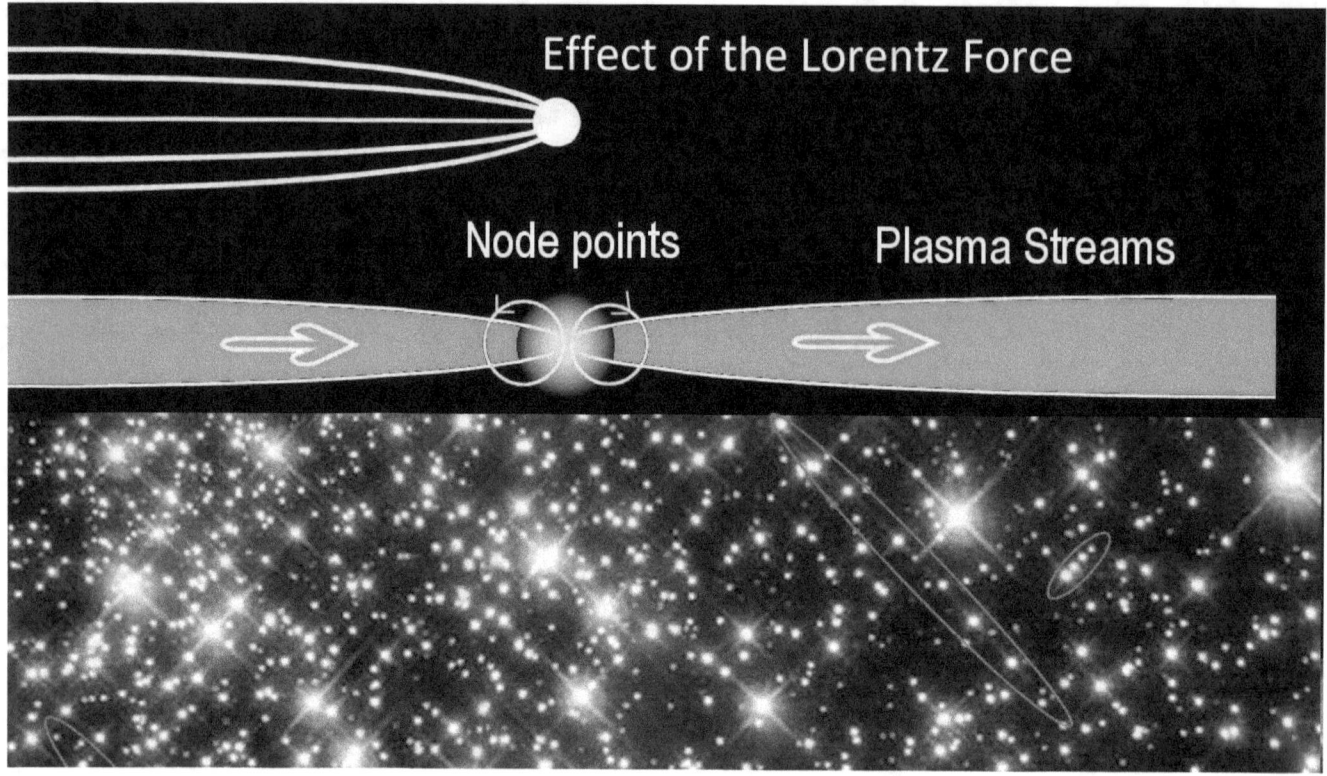

At the node point something interesting happens. When plasma is in motion in the same direction, the electric movement creates magnetic fields that pull the plasma together into ever-smaller cross-sections, which in turn increases the 'pinching' effect, and with it the magnetic field strength.

Dynamic electromagnetic structure, the Primer Fields

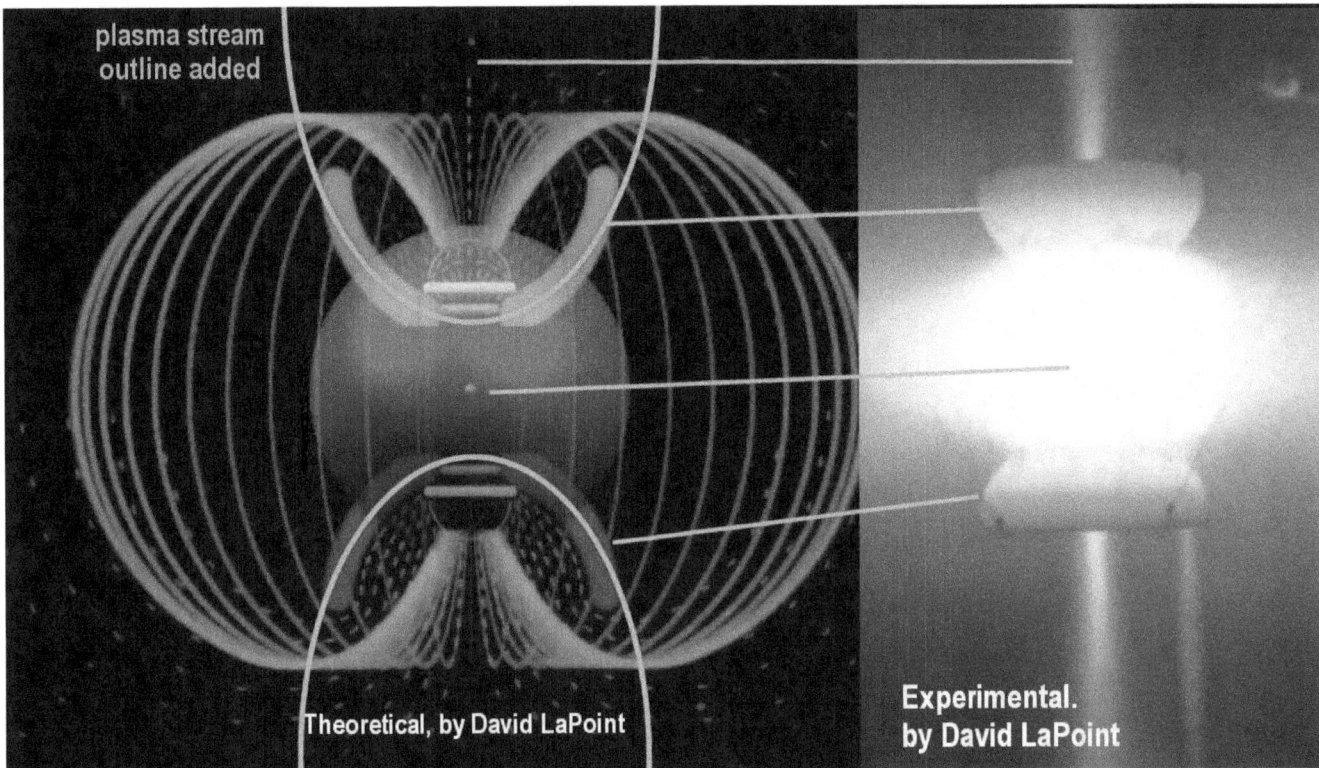

At an extreme point the entire process breaks down, whereby a hole opens up in the plasma stream, through which intensely concentrated plasma escapes. That's where we find our Sun located. The escaping plasma is literally focused onto our Sun by the remaining intense magnetic fields that remain active. The plasma researcher David LaPoint, who explored the process, has termed the dynamic electromagnetic structure, the Primer Fields.

When the Primer Fields are active

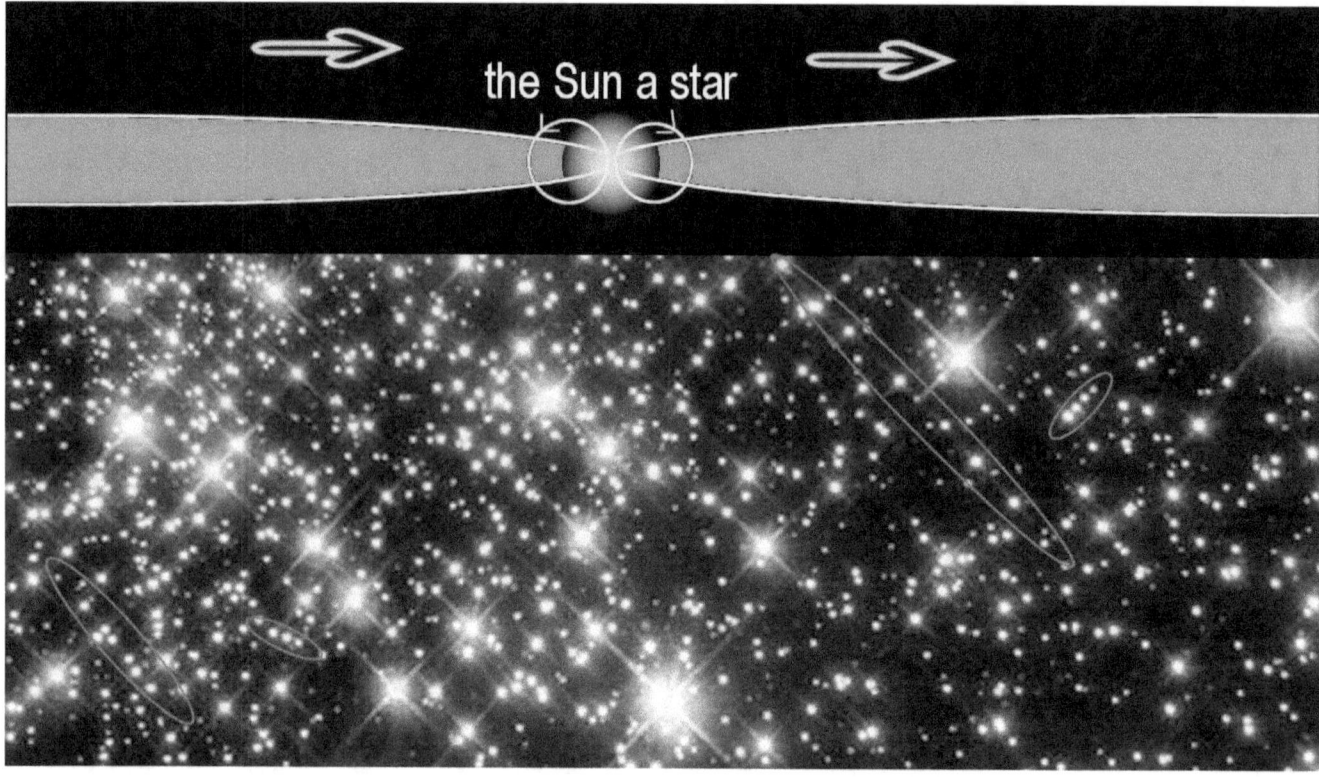

When the Primer Fields are active, our Sun is intensively powered, with a dense sphere of plasma surrounding it, which is formed by the Primer Fields.

When the Sun hibernates

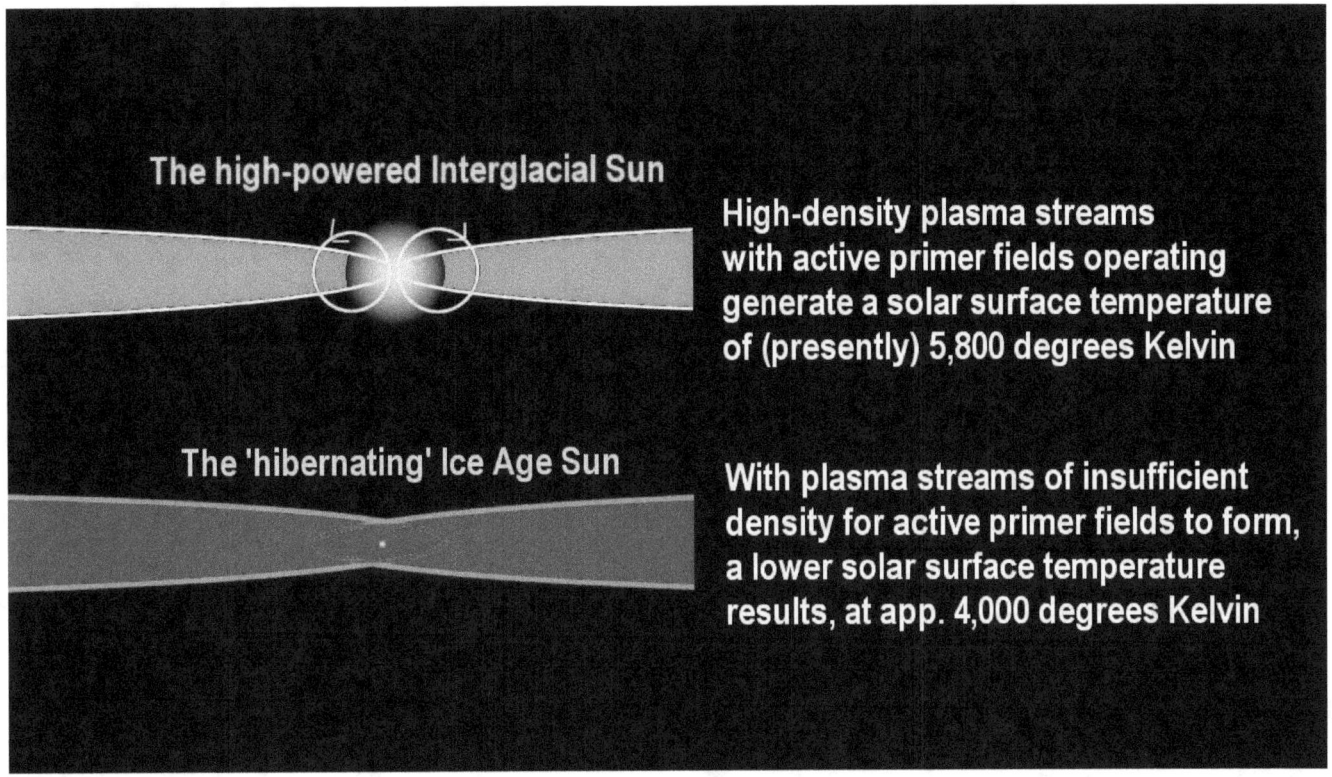

But when the interstellar plasma stream is too weak to maintain the Sun in its high-powered state, the pinching effect is too feeble for the Primer Fields to form, so that the plasma simply flows by the Sun. When this happens, the Sun operates at a dramatically lower intensity. It goes to sleep. It hibernates, waiting for the plasma streams to become dense again.

When the Sun hibernates, the Earth is in an Ice Age climate.

Hibernating for 85% of the time

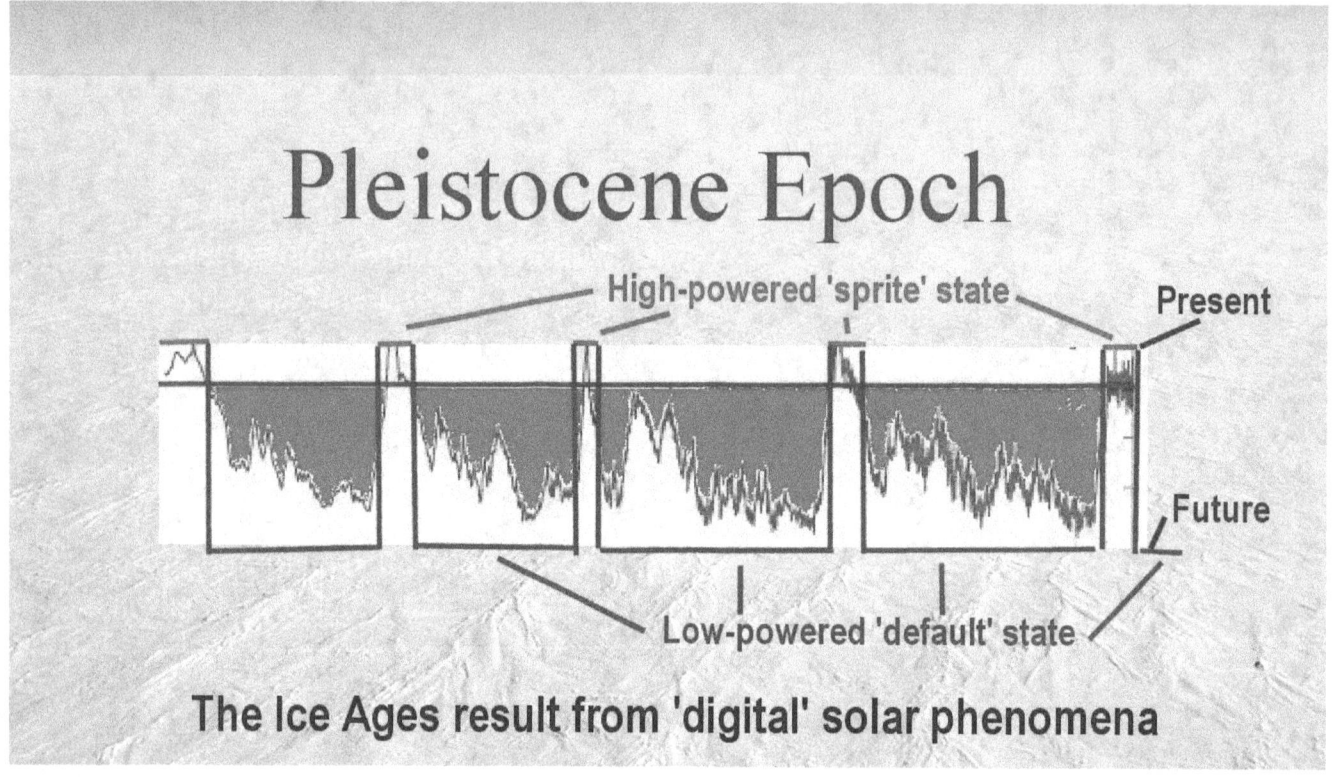

In the long sweep of the last half-million years, the Sun has been hibernating for 85% of the time.

Active with Primer Fields

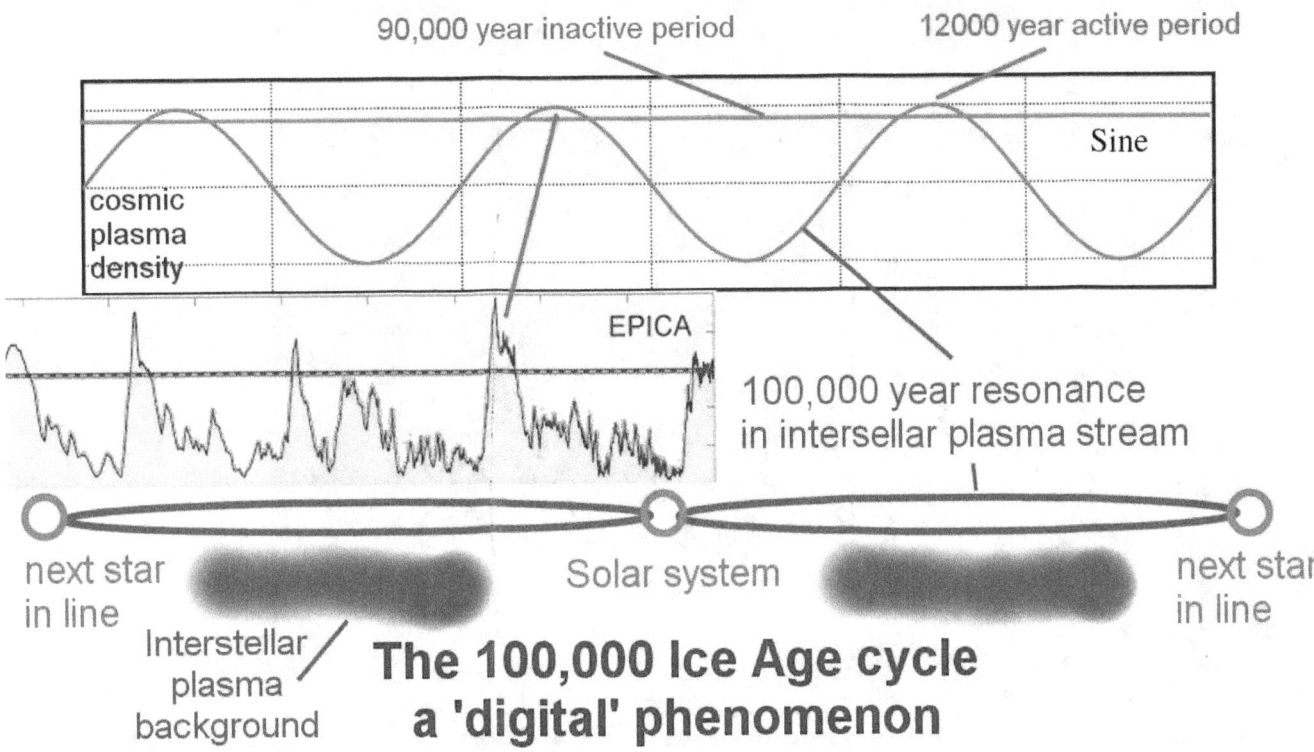

The hibernation of the Sun is the result of the present low density in the interstellar plasma stream and its internal oscillation. There is only enough density left at the present epoch for the Sun to become "Active with Primer Fields" for a mere 15% of the time at the peak of the oscillation.

The steep decline, near the off-point

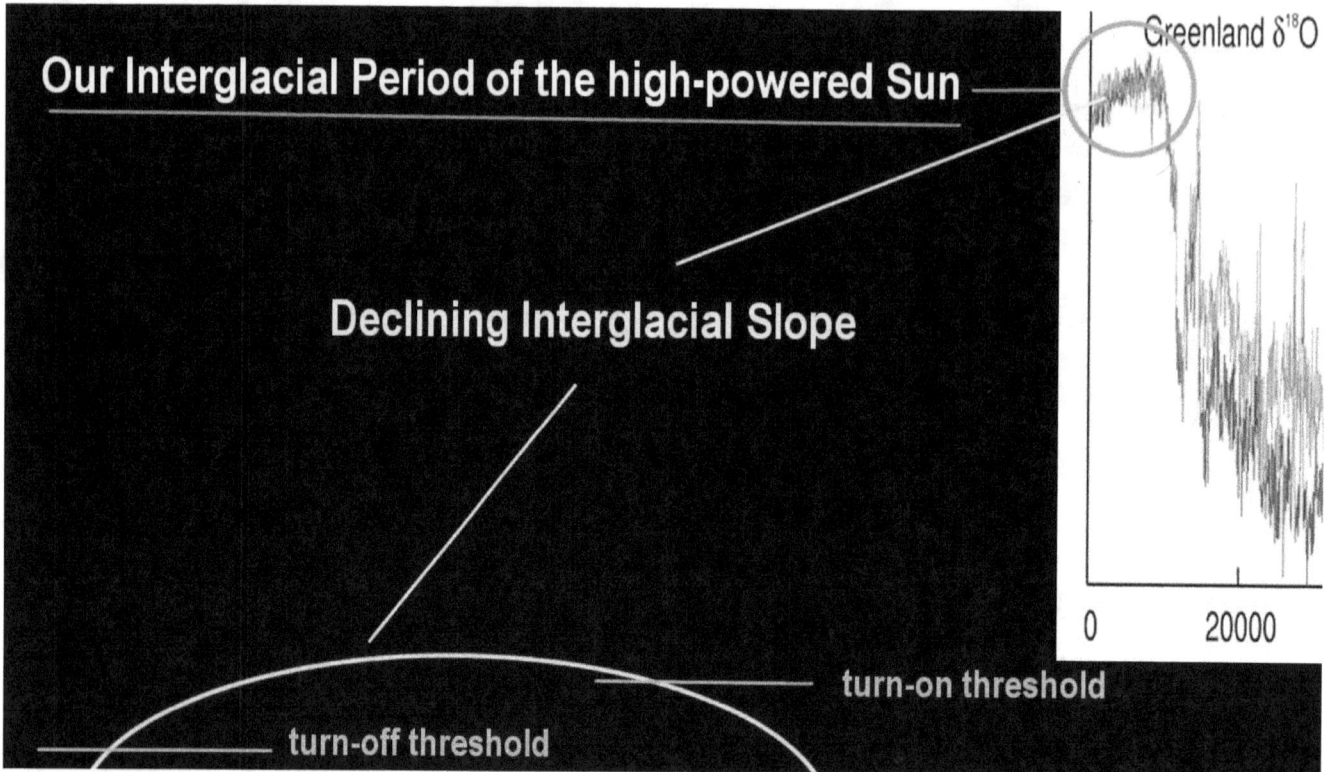

The peak-time activation is the reason why the interglacial period has been diminishing from its mid-time to the present.

As I presented in a previous video, it takes a far-greater plasma-density to activate the Primer Fields, then it takes to maintain them.

Note that the plasma density diminishes evermore steeply, the closer the interglacial period comes to the turn-off threshold for the Primer Fields. The steep decline, near the off-point, is the DRIVER for all the diminishing processes that we see now expressed evermore in the weakening of the solar system.

The diminishing yellow curve of the interstellar plasma density that is shown here, between the on and off states, has a direct effect on the intensity of the activity of our Sun during the interglacial period, and especially so now, as we come near the end of the interglacial. The on/off points on the curve mark the extend of interglacial period.

With this established, we can now look at the details of the diminishing solar system and develop a model that can facilitate the kind of phenomena that we see.

Three types of different resonance effects

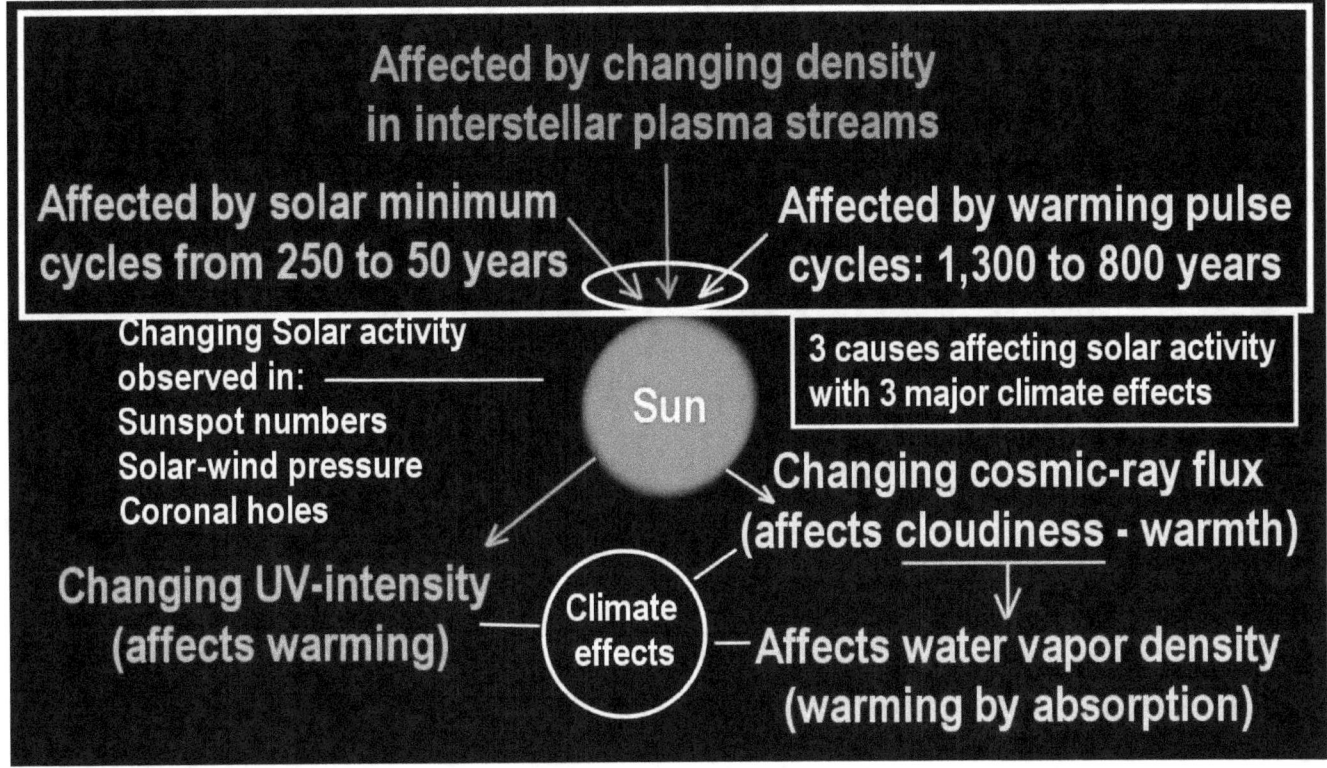

The intensity of the solar actions that we see evident in historic ice core samples appears to be affected by three types of different resonance effects, which flow together and affect the density of the plasma streams flowing onto the Sun. This means that the three different resonance effects originate from different plasma structures.

We see for example that the solar minimum cycles are of shorter duration than the intervals of the big historic warming pulses, which therefore must both have a different cause. This means that the specific climate features that we see reflected in the solar system's dynamics, each add their unique contribution to the interstellar plasma stream, with all of the contributions flowing together in the process that powers the Sun.

➢ **The innermost resonance is slowing**

> The innermost resonance,
> the pulse of the solar system,
> the 11-year solar cycle,
> **is slowing down**

The innermost resonance, the pulse of the solar system, the 11-year solar cycle, is slowing down.

The cycle time has become longer

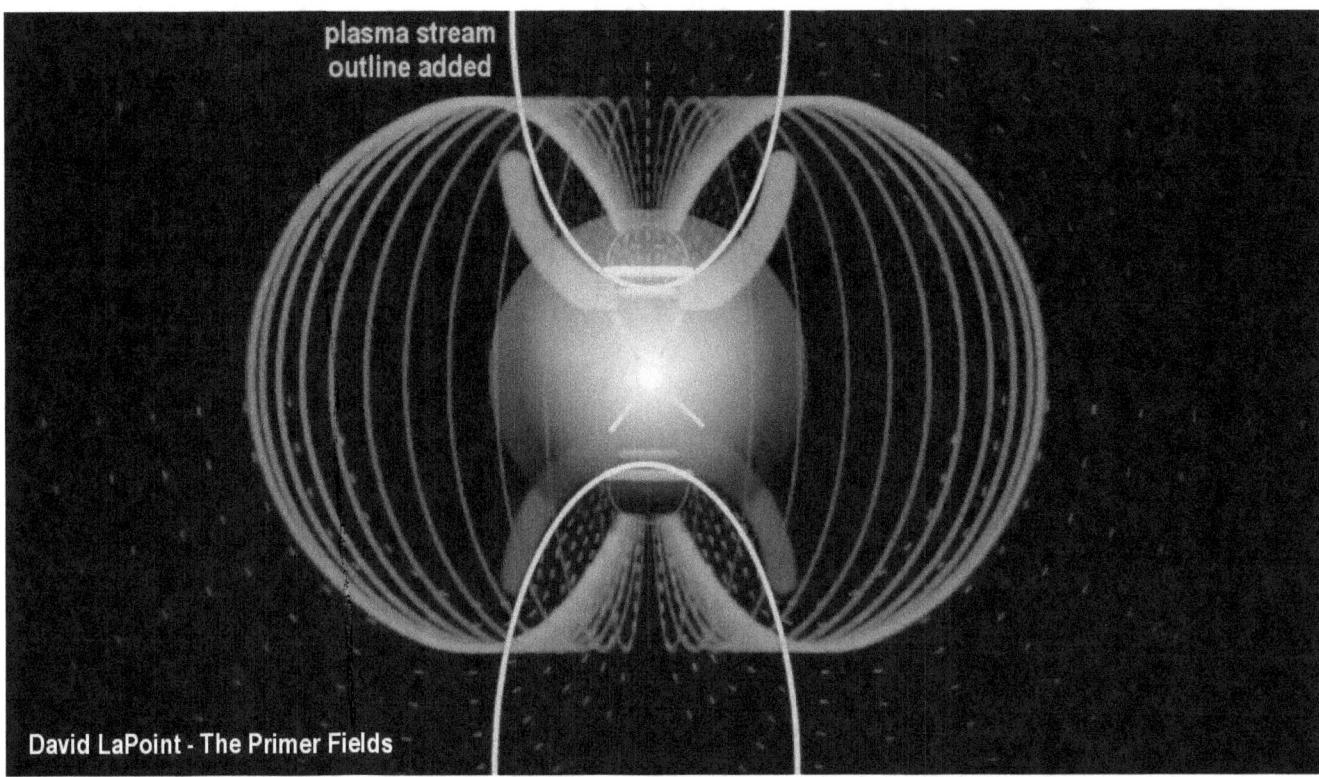

The researcher David LaPoint discovered in laboratory experiments, that when the Sun is located between two symmetric electromagnetic structures, then the lightest imbalance between the two structures causes a magnetic polarity to be imposed onto the Sun, according to the imbalance.

When the entire plasma structure that combines the Primer Fields is resonating, the imbalance resonates with it. As the structure resonates, it imposes alternating magnetic polarity unto the Sun. With an 11-year resonance built into the system, the resonance cycles cause the Sun's magnetic polarity to flip every eleven years, for a complete cycle spanning 22 years.

The solar activity intensity fluctuates during the course of each cycle, though with the cycle time remaining constant. The magnetic polarity flips during the 'low' intensity portion of each solar cycle.

Only in very recent time, from the Year 2,000 onward, has the cycle time has become longer.

Now this last bastion of stability is diminishing

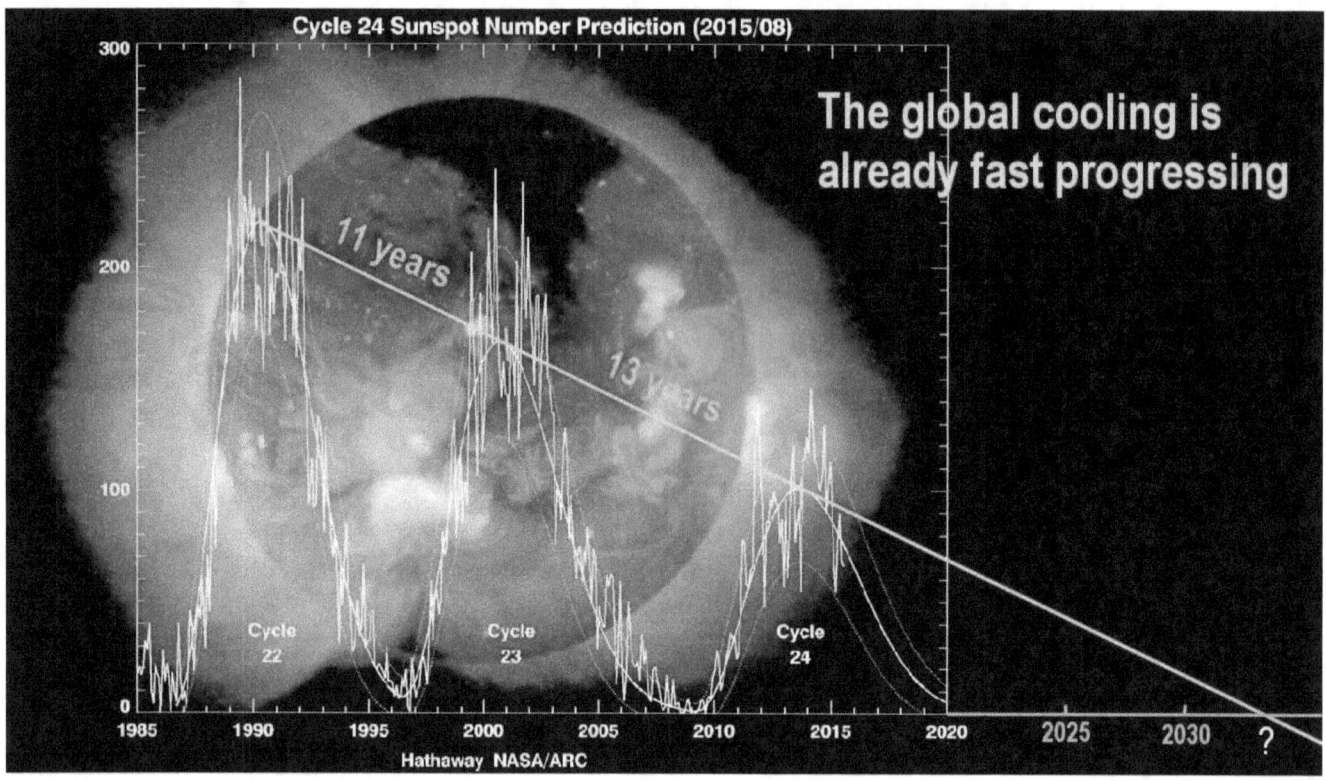

During the interval between the peak of cycle 23 and cycle 24, the cycle time became extended to 13 years. This is a totally new phenomenon, without an equal for centuries.

Until the year 2000, the timing of the solar cycle had remained the only stable element of the solar system. It had weathered the sweeping consequences of the weakening Sun that has affected everything else. Now this last bastion of stability is diminishing too, and at a tremendous rate of change.

Of course, the Sun itself didn't cause the sudden change of pace.

The Primer Fields appear to have shifted

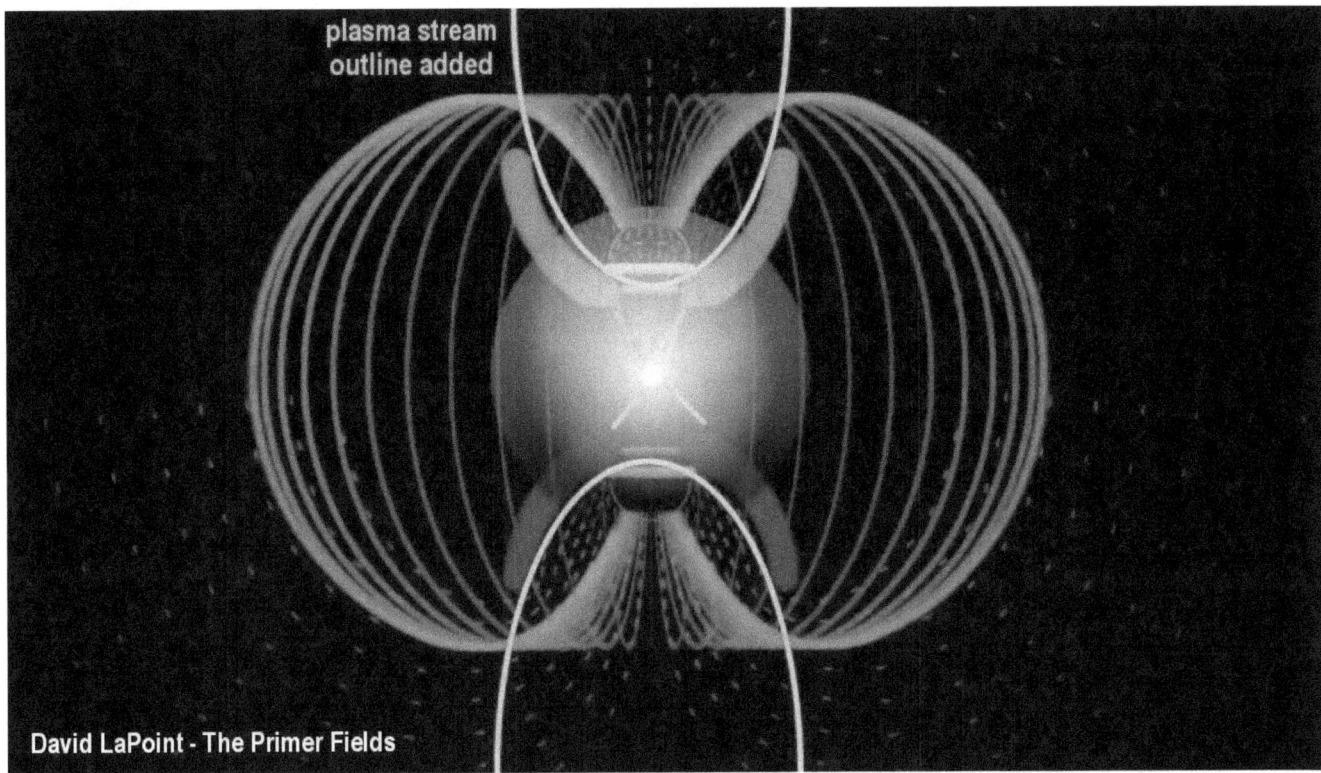

The positions of the Primer Fields appear to have shifted. The Primer Fields appear to have receded further away from each other as the result of the entire solar system becoming weaker. The increased space enabled the plasma structure between the Primer Fields to expand. As the result of it, the larger plasma structure produces longer electric resonance cycles, and consequently longer solar cycles.

This increase happened dramatically fast

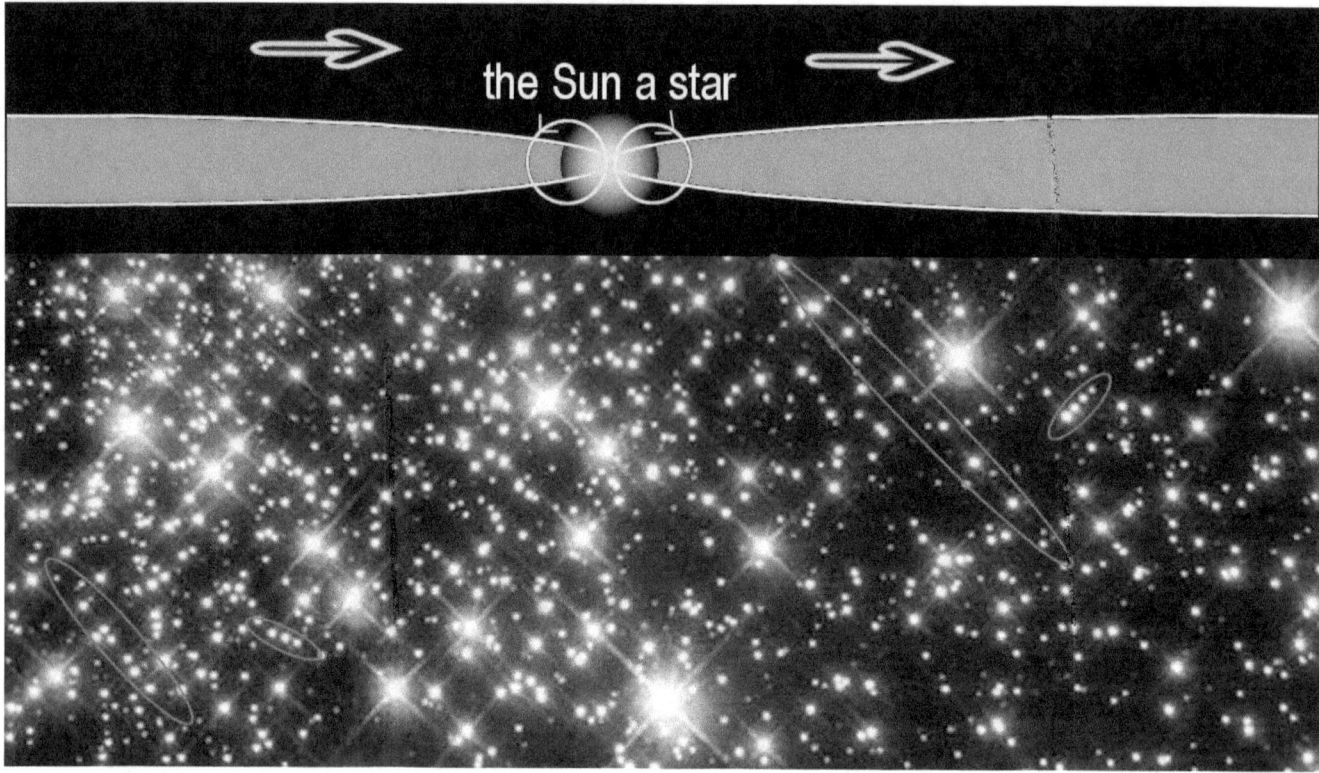

And this increase in the spacing of the node point didn't happen spasmodically. It happened dramatically fast.

From 11 years to 13 years This is big!

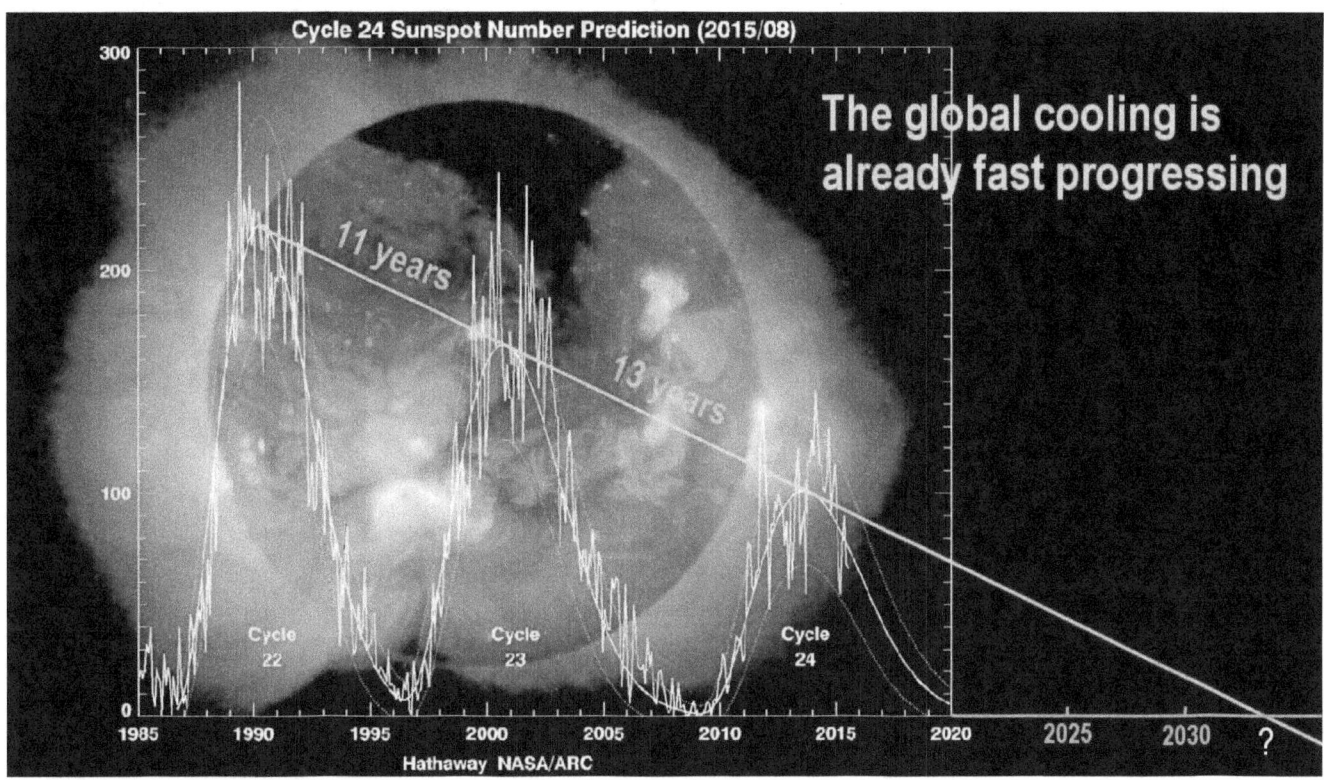

The increase of the cycle time from 11 years to 13 years amounts to a whopping 18% expansion over the span of roughly a decade. Wow! This is big!

Extreme uncertainties laying ahead

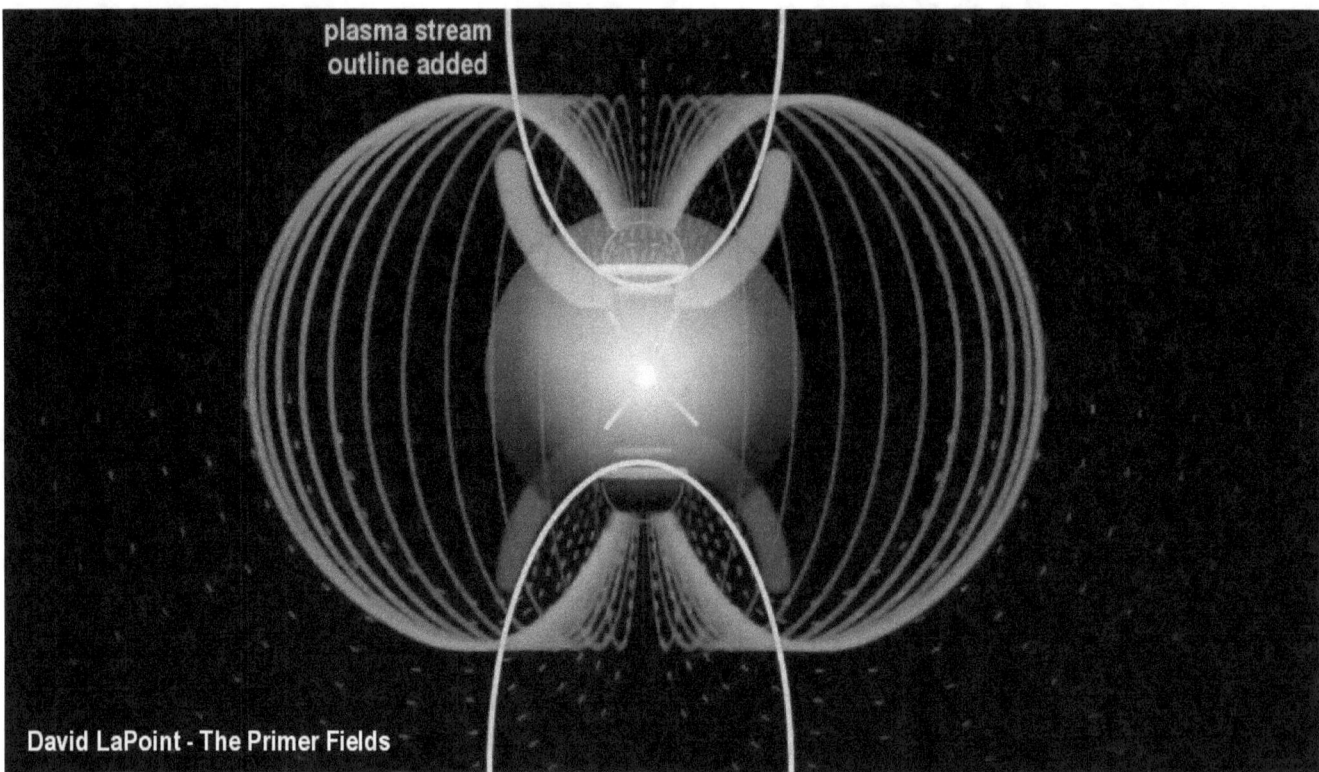

If this can happen to the most solid structure in the solar system, and in just a single decade, we are looking at extreme uncertainties laying ahead of us.

The entire solar system has now become so weak that the Primer Fields are beginning to shift, and shift rapidly, so that the very heart of the solar system, the beat of the 11-year solar cycle is running slower.

Deep climate changes are beginning to unfold

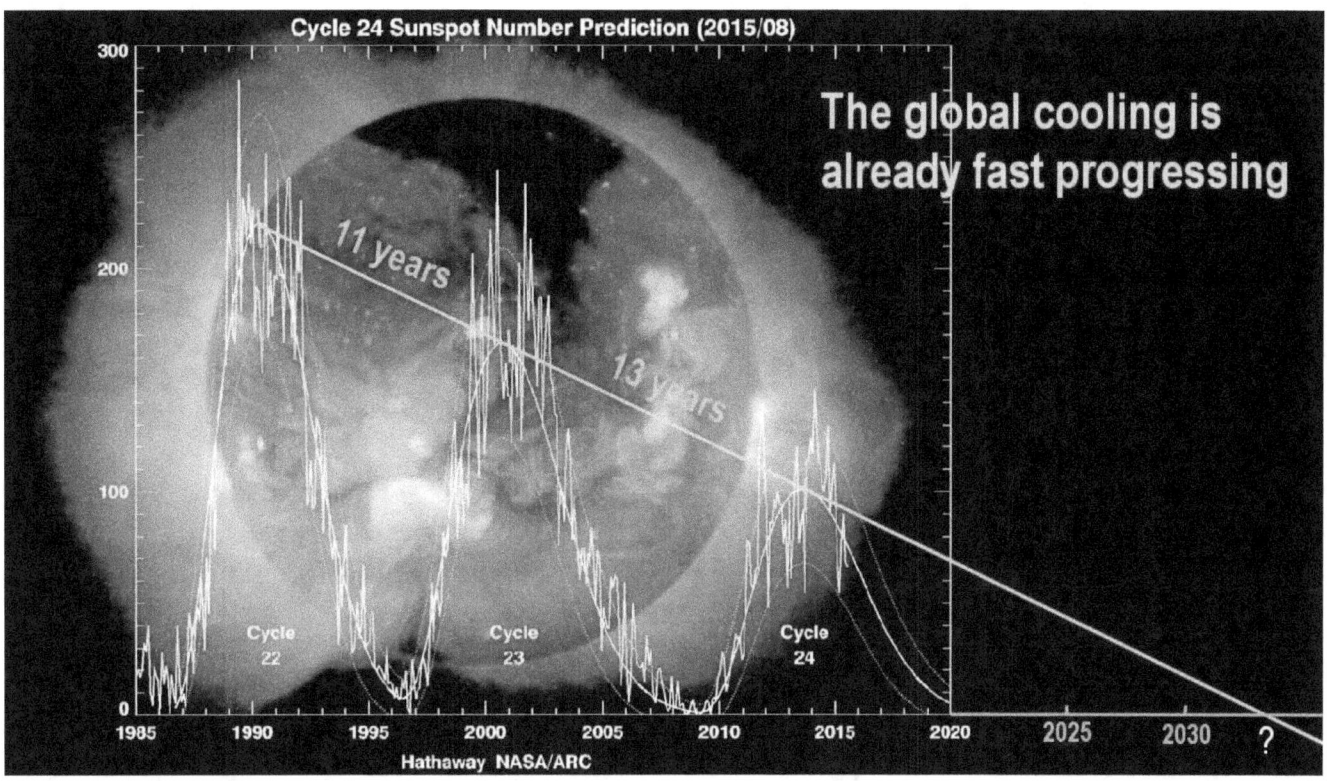

Against this background everything becomes unpredictable. It is not surprising, therefore, that deep climate changes are already beginning to unfold.

The changing long cycles

> The changing long cycles

The changing long cycles.

Deep changes for 3,000 years already

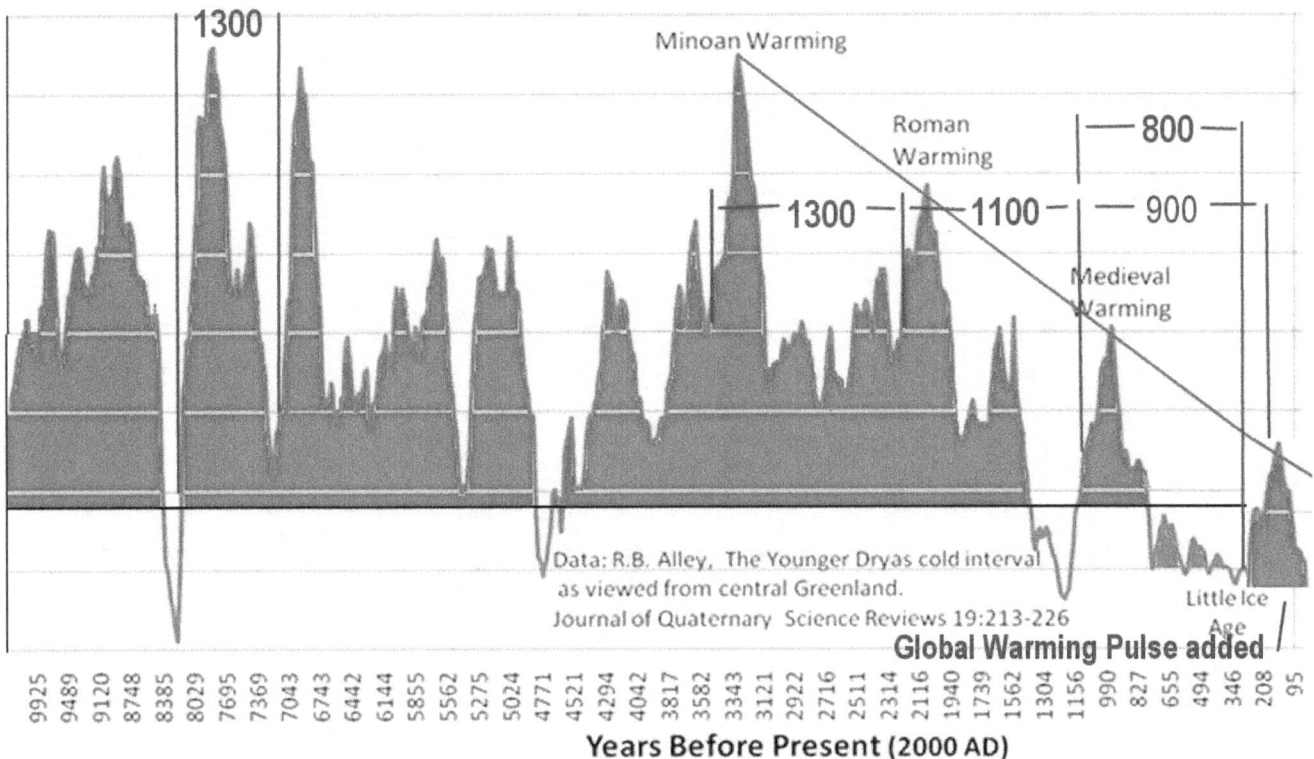

As I said before, deep changes have begun to manifest themselves for 3,000 years already, as we see it reflected in the diminishing big warming pulses and the shortening of their cycle-intervals.

Grand Solar Minimum

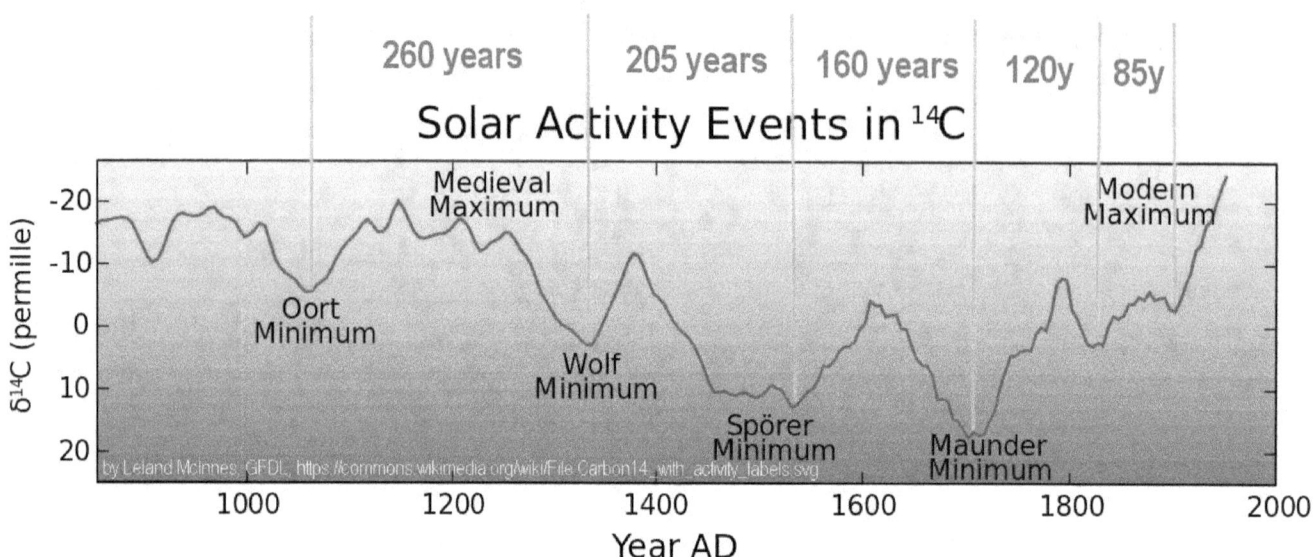

As I also said before, we see the big Grand Solar Minimum events occurring in ever-shorter intervals and amplitude.

The 250-year cycles getting shorter

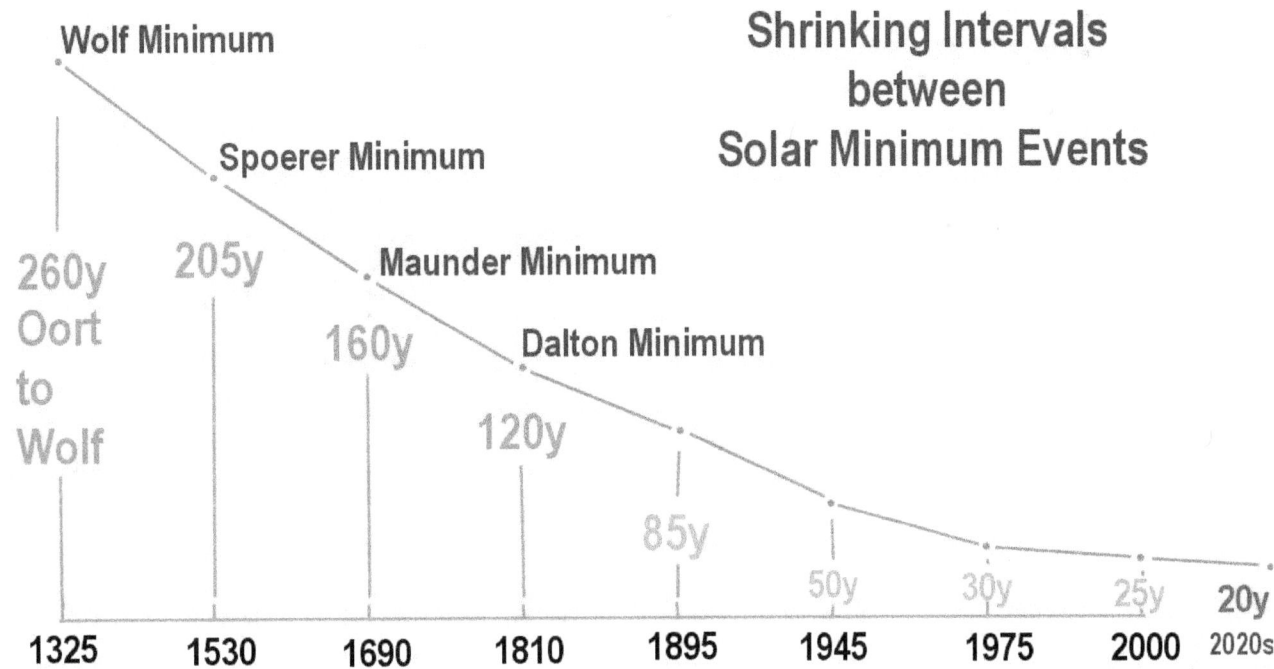

We see the 250-year cycles getting shorter.

The 1,300-year intervals getting shorter

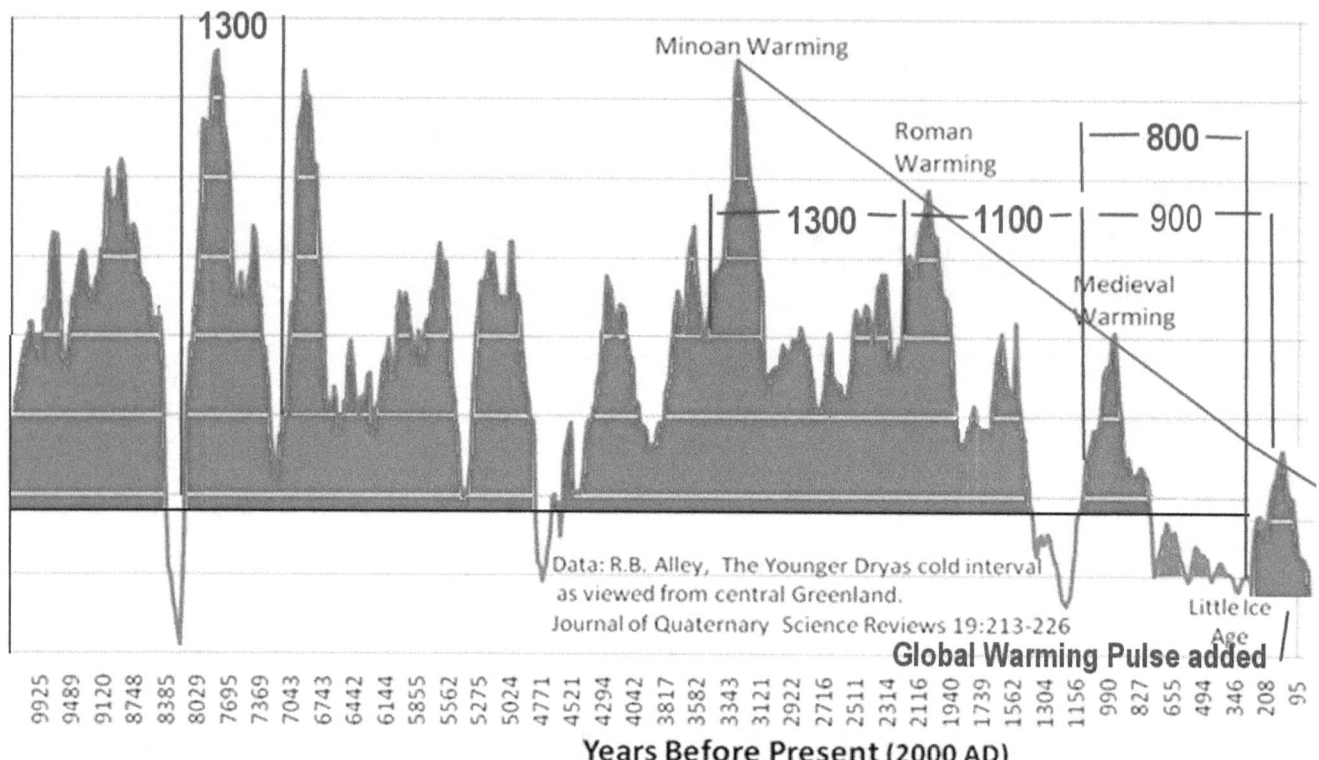

And we see the 1,300-year intervals getting shorter likewise.

Both of these long-term cyclical phenomena are rapidly diminishing.

Caused by separate phenomena

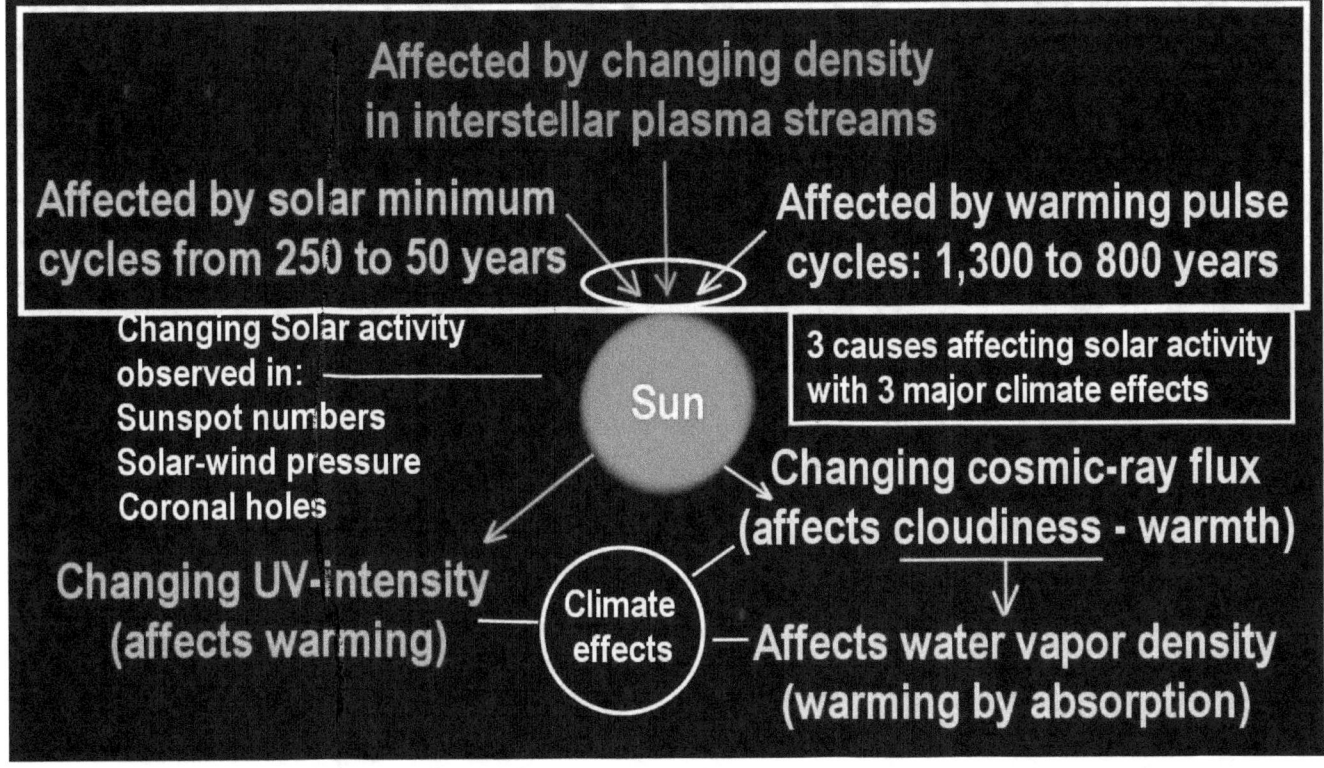

However, the big difference in basic cycle times, of 1,300 years versus 250 years, indicate that these events are caused by separate phenomena, though of a similar style, and of a type that affects what flows through the Primer Fields onto the Sun.

Of two-fold nested Primer Fields

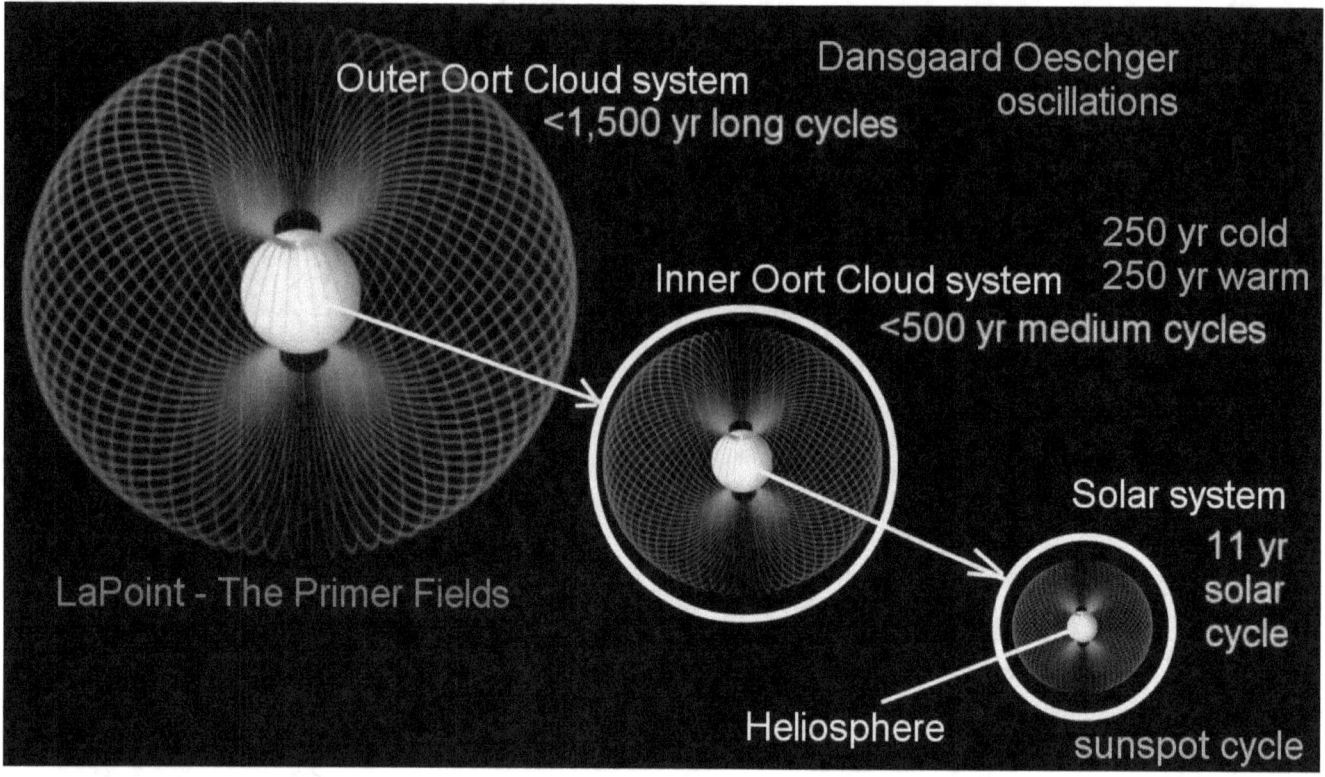

This dual-type evidence leads to the concept of two-fold nested electromagnetic structures, or nested Primer Fields, as David LaPoint refers to them, a concept that's similar in geometry of the theorized Oort Cloud system.

The Oort Cloud of nested asteroid fields

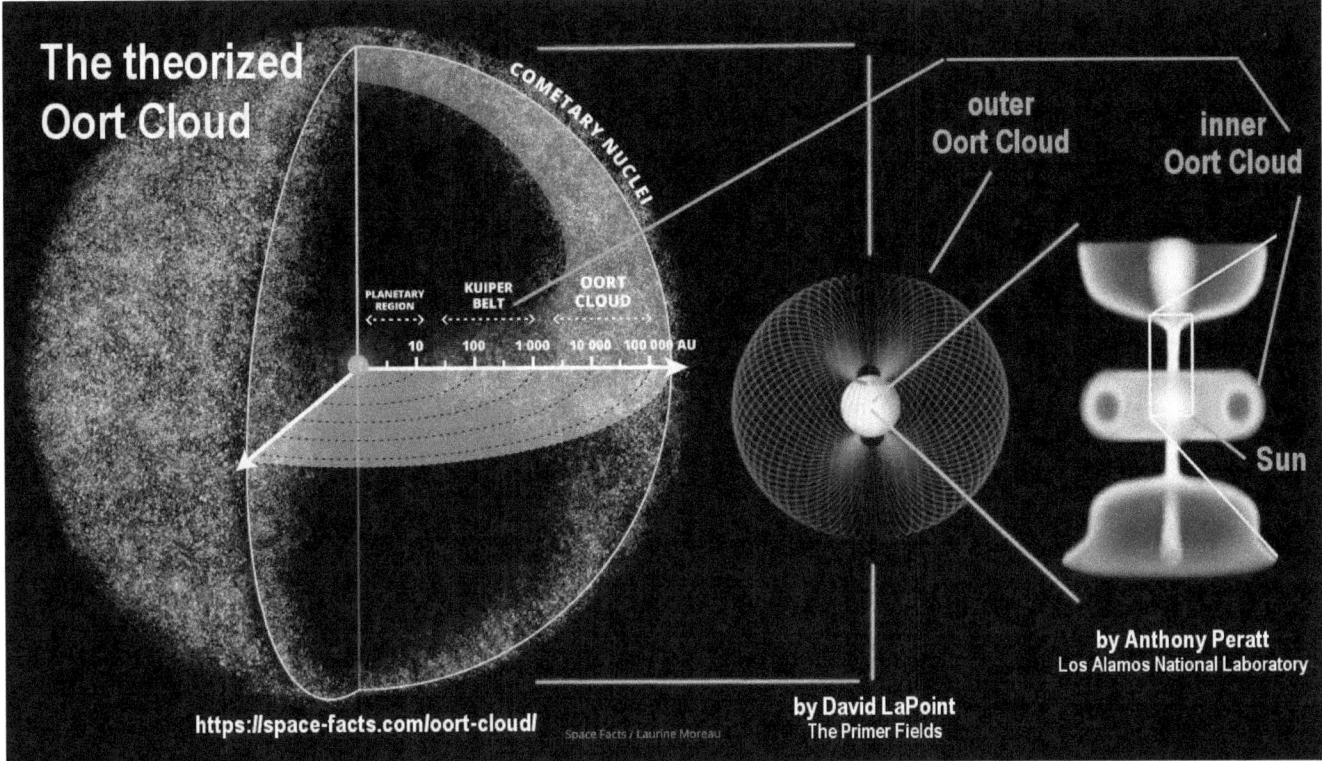

The Oort Cloud is a concept of nested asteroid fields that are theorized to surround the solar system at a large distance. The Oort cloud was theorized as a potential origin for comets that are not aligned with the planetary ecliptic. The cloud is theorized to contain trillions of asteroid type objects, all held in place in two concentric clouds, a large shell-type cloud, and a toroidial cloud within it that is sometimes referred to as the Kuiper Belt. Both structures are perceived as centered on the Sun, but at a very long distance away from it.

However, the theorized clouds of asteroids, may be simply plasma structures.

The geometry of such structures is consistent with features discovered in laboratory experiments of high-energy plasma physics, shown on the right in pink. In a high-energy electric discharge experiment conducted at the Los Alamos National Laboratory. The experimental plasma stream became extremely concentrated by the self-forming magnetic Primer fields, into a narrow super-dense stream of plasma. Since the experiment did not incorporate a sink for the plasma as in the case of the Sun in cosmic space, the super-dense plasma stream created its own energy dissipation by forming a toroid of energized plasma around it. The dissipation enabled the now weaker stream to expand again in the reverse of the process.

In comparison, the plasma toroid would be comparable to the theorized Kuiper Belt of the Oort Cloud structure.

The researcher David LaPoint discovered that the opposite magnetic polarity of the two primer fields, creates a large magnetic field that encapsulates the entire structure. This is what the theory of the outer Oort Cloud may represent.

The physical asteroid objects that supposedly compose the Oort clouds, have never been actually seen. Their existence has been assumed as a potential source for comets. This means that the theorized clouds of asteroids may not actually exist as clouds of asteroids, but may exist instead as plasma structures which do have an effect on comets passing through them.

Nested plasma structures are possible

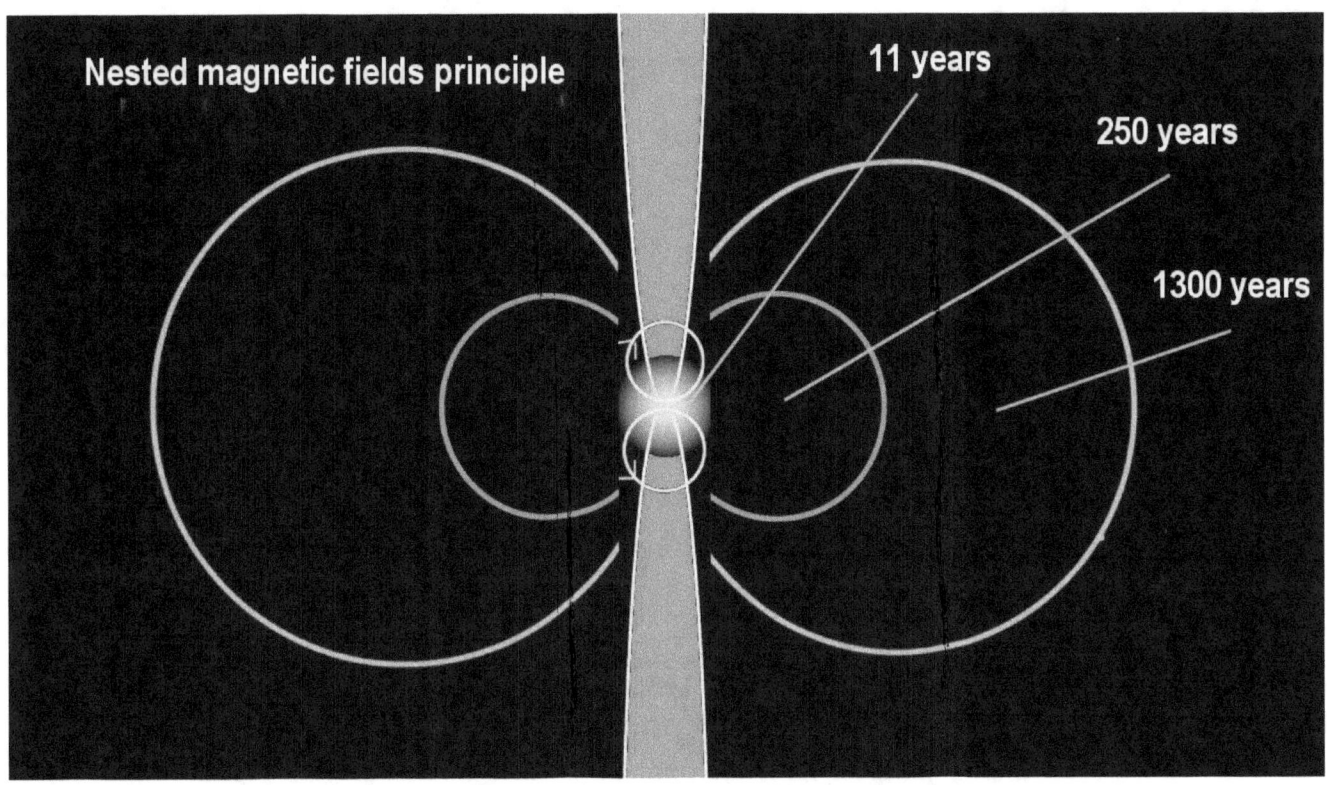

In principle, nested plasma structures are possible that echo the basic nested concept of the theorized Oort cloud.'

The toroidial shape

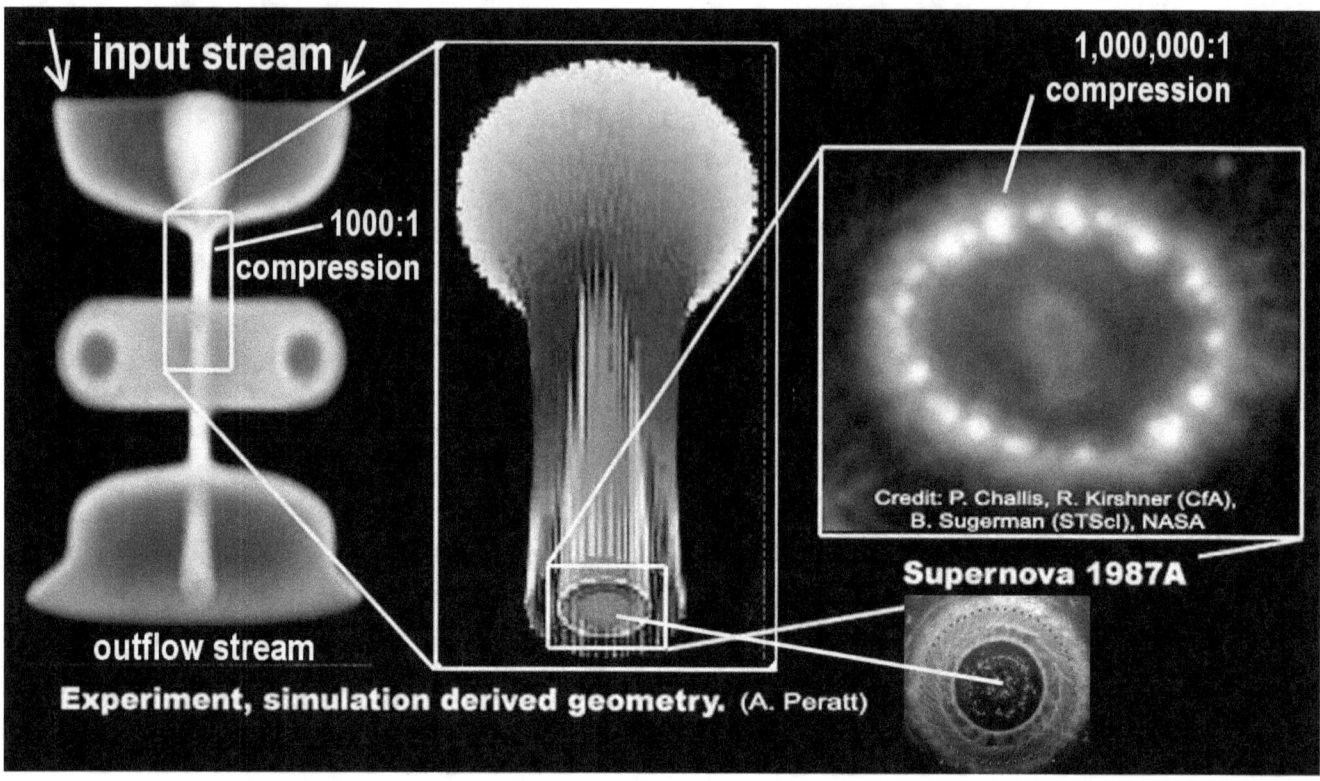

The toroidial shape that developed around the central stream in the high-energy plasma experiment at the Los Alamos National Laboratory, illustrates a basic principle that may be reflected in the forming of the larger cosmic plasma structures.

As giant plasma structures

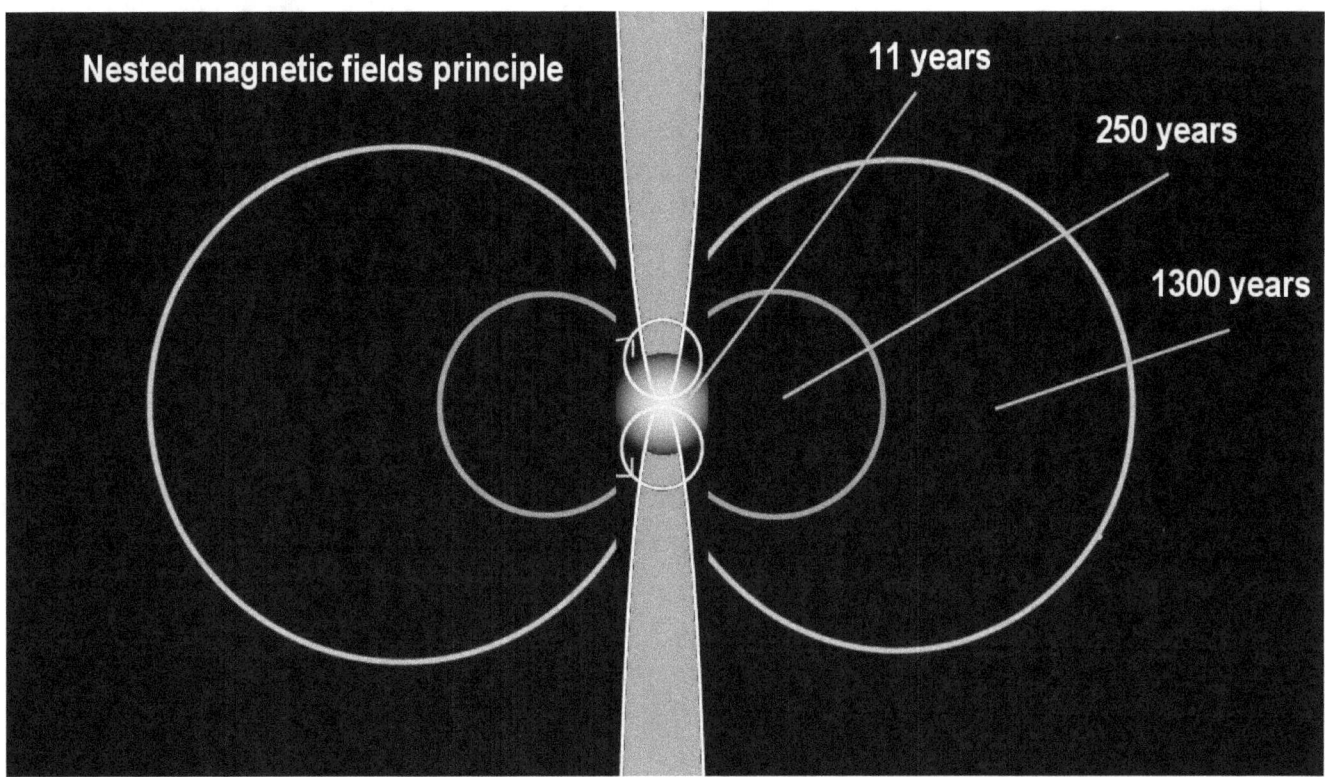

Of course, as giant plasma structures, each of the big structures has its own inherent electric resonance that reflect its physical size.

This means that the resonance in the outer plasma cloud is likely the cause for the 1,300 years cycle of the big warming pulses that we see in climate records, and the resonance of the smaller, toroidial inner plasma cloud could be the cause for the 250-year cycles of the Grand Solar Minimum periods.

Since these plasma structures that are a part of the solar system, are dynamically formed, their size can also dynamically vary. As the solar system is now weakening, the physical size of these structures would therefore become correspondingly smaller, which would reflect itself in shorter resonance times, such as we see in the ice core records.

The 250-year resonance cycle became faster

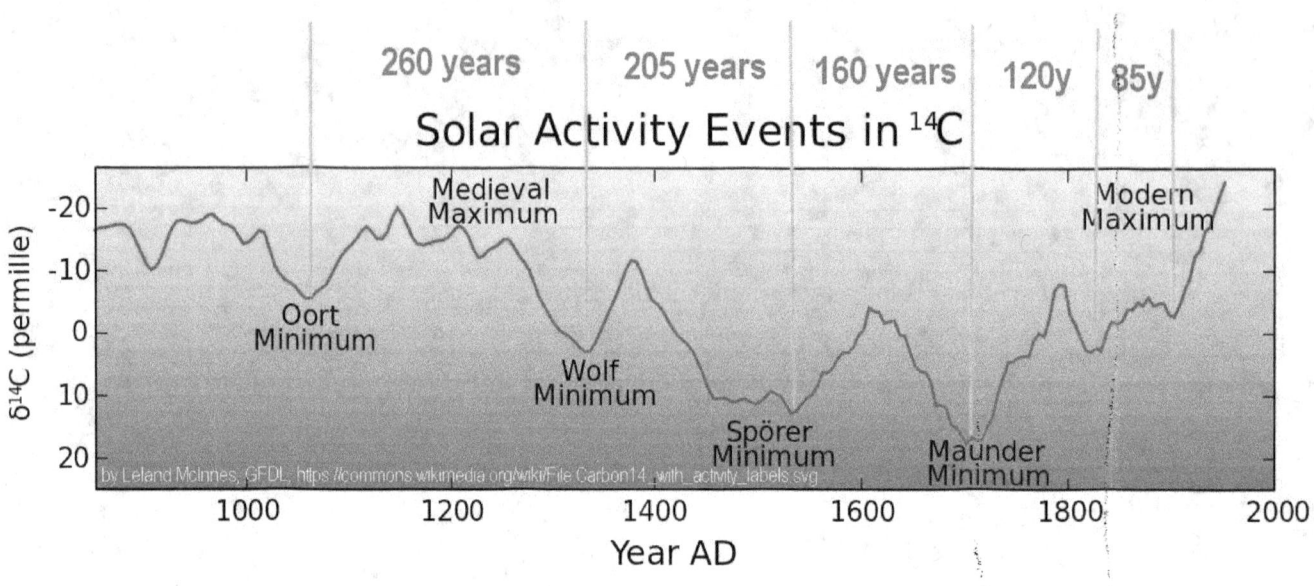

The 250-year resonance cycle, thereby became faster. It is repeating now in ever shorter intervals.

An orderly geometric regression

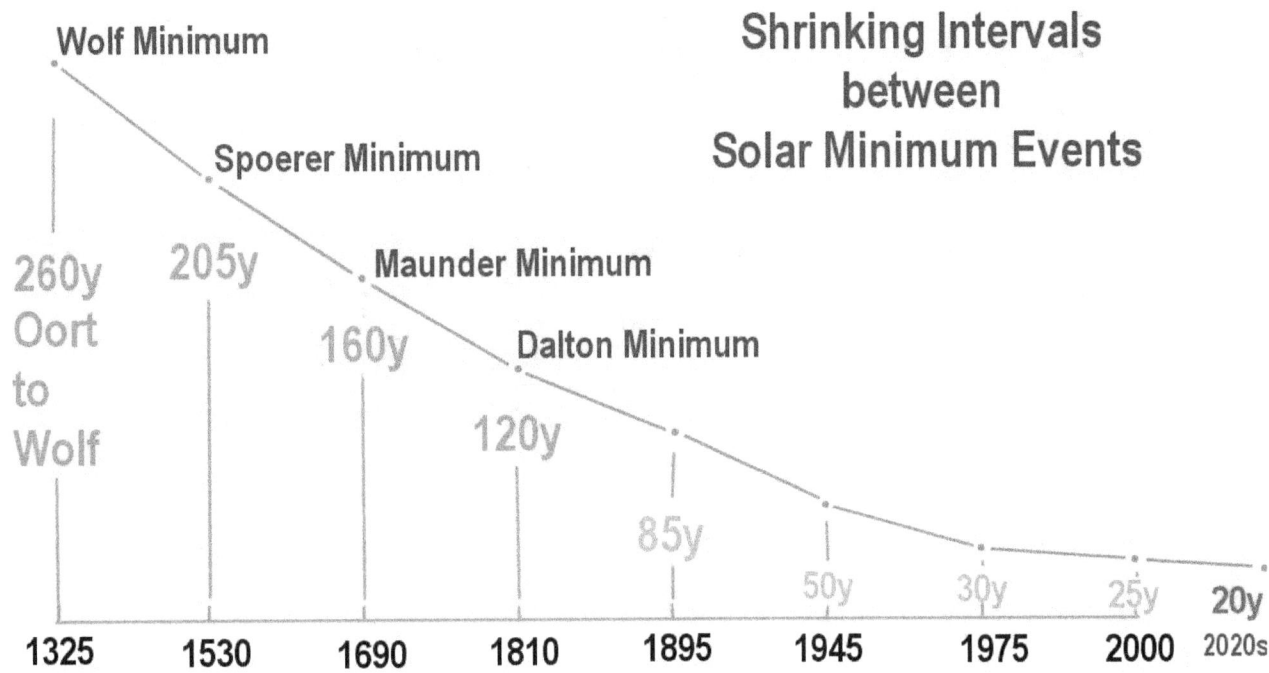

The measured regression, as we see it expressed in ice core samples, unfolds as an orderly geometric regression for which the outcome is somewhat predictable.

The 1,300-year cycle is now beating faster

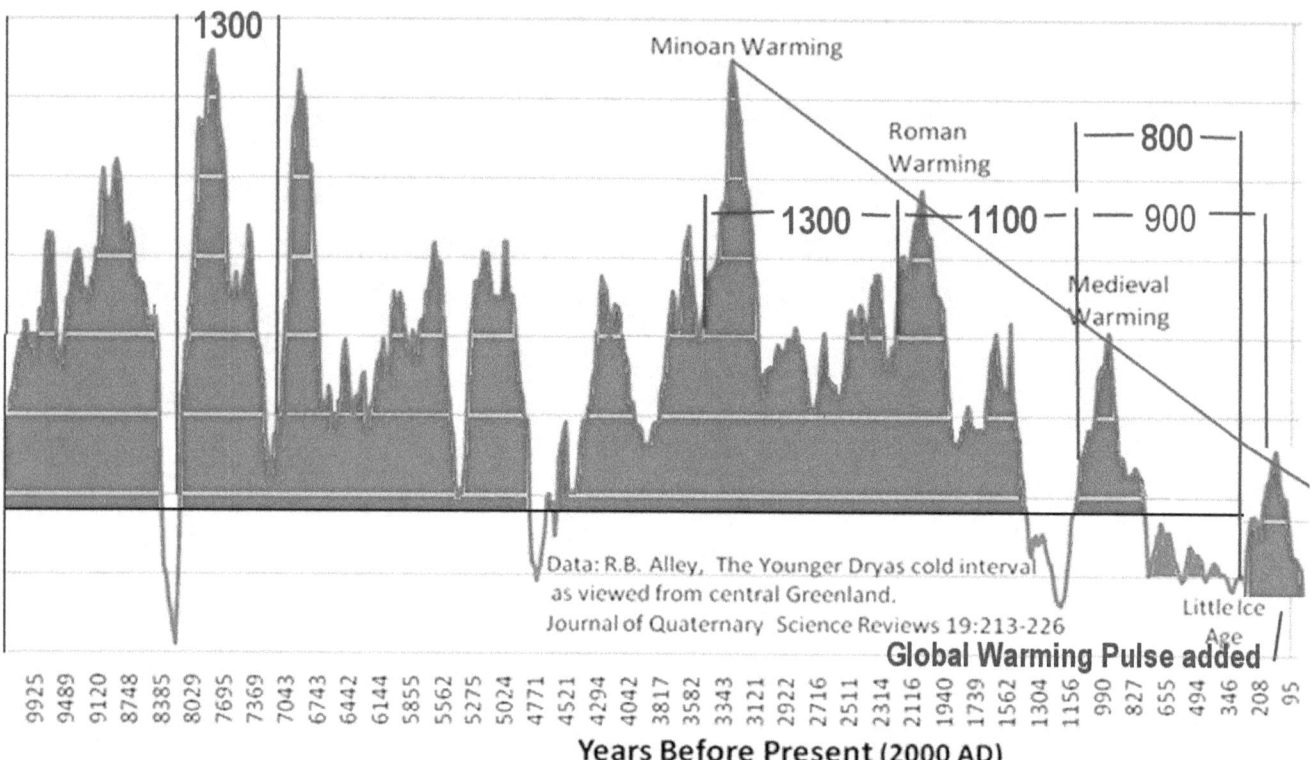

Likewise, the 1,300-year cycle is now beating faster. Its repetition rate was at its last appearance only 800 years in length.

This shrinking cycle time may have saved us in the 1700s from the Little Ice Age becoming the Big Ice Age.

➤ In the sub-visible domain

> In the sub-visible domain

In the sub-visible domain.

The cyclical Warming spike in the 1700s

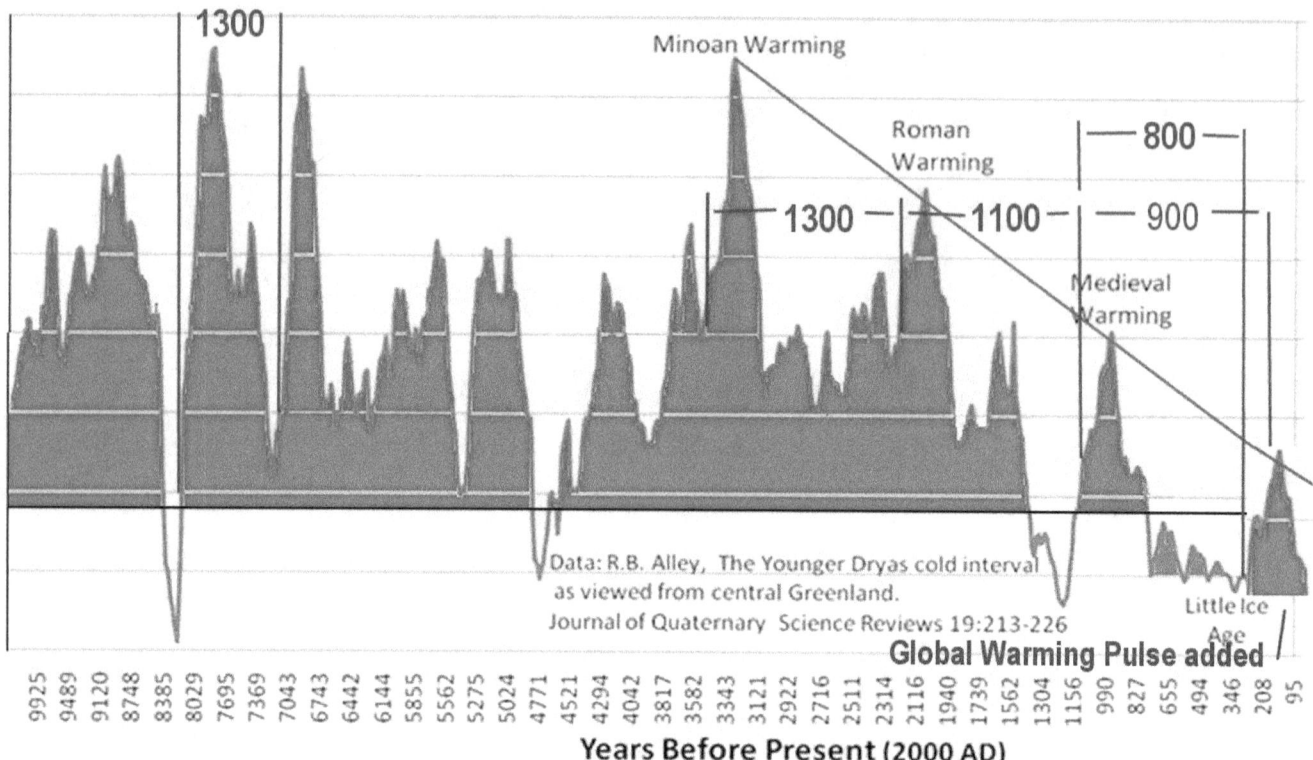

The cyclical Warming spike in the 1700s, though diminished somewhat, had up-ramped the Sun at the deepest point of the Little Ice Age, and with it may have saved our existence.

No one was prepared at the time

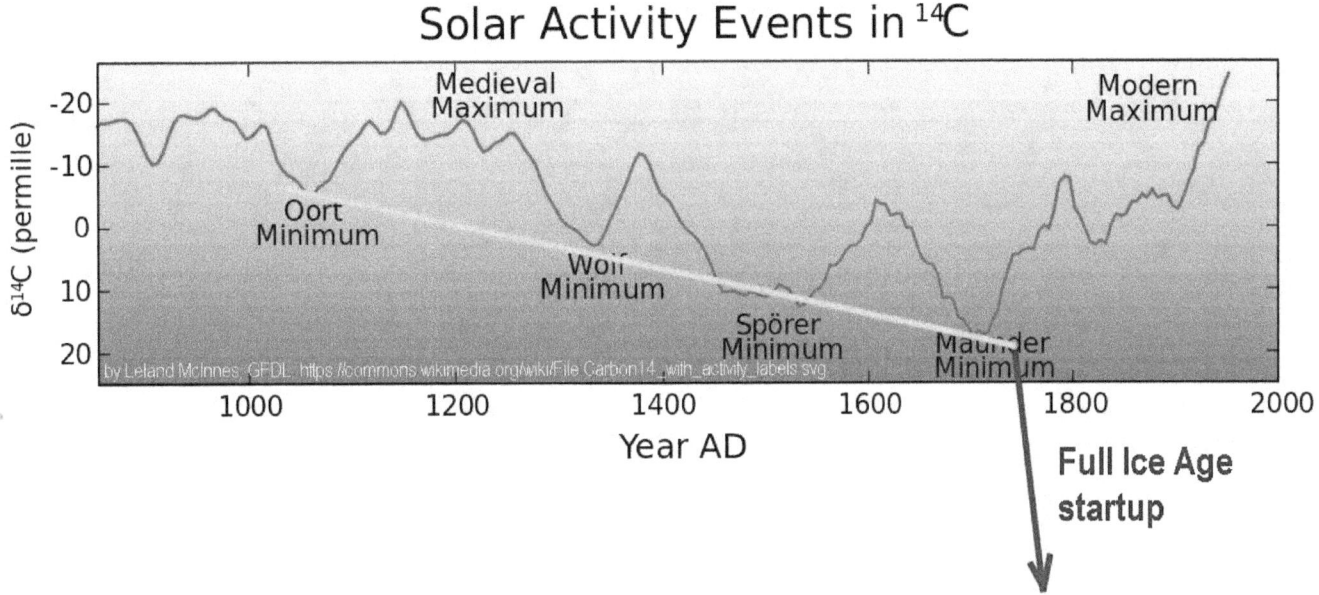

No one was prepared at the time for living on a largely uninhabitable planet, which the Earth becomes during the Ice Age glaciation phase.

All through the 1600s

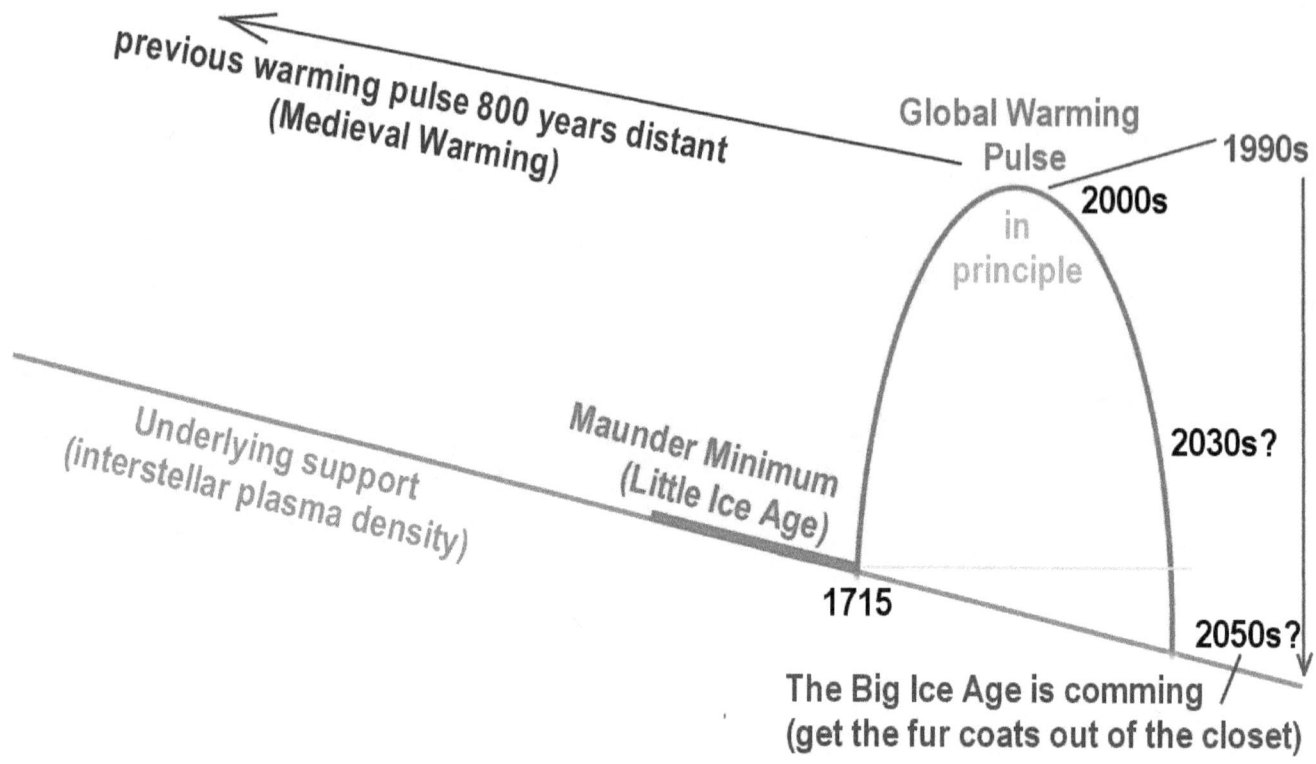

All through the 1600s, the background level for the Sun was so weak that almost no sunspots were visible for three decades, and only a very few had been seen during the decades before. The actively lean period became known as the Maunder Minimum, and is famous for its lack of sunspots.

Did the solar cycles stop?

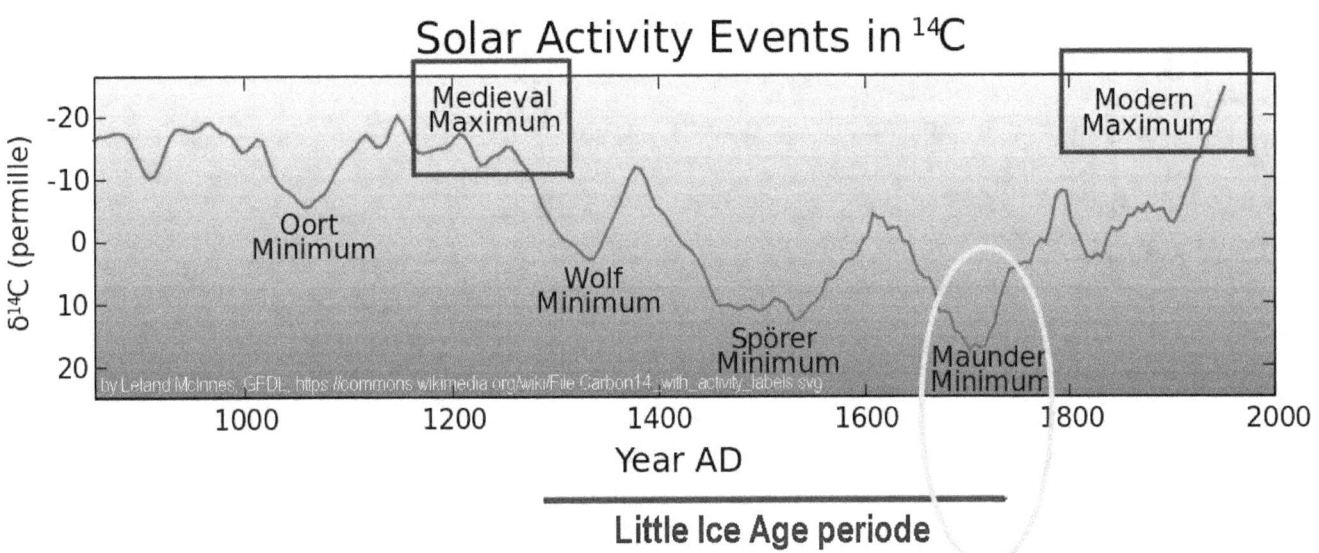

But what about it? Did the solar cycles stop? Did the Sun go to sleep during the Maunder Minimum?

No, the Sun didn't go to sleep during the Maunder Minimum

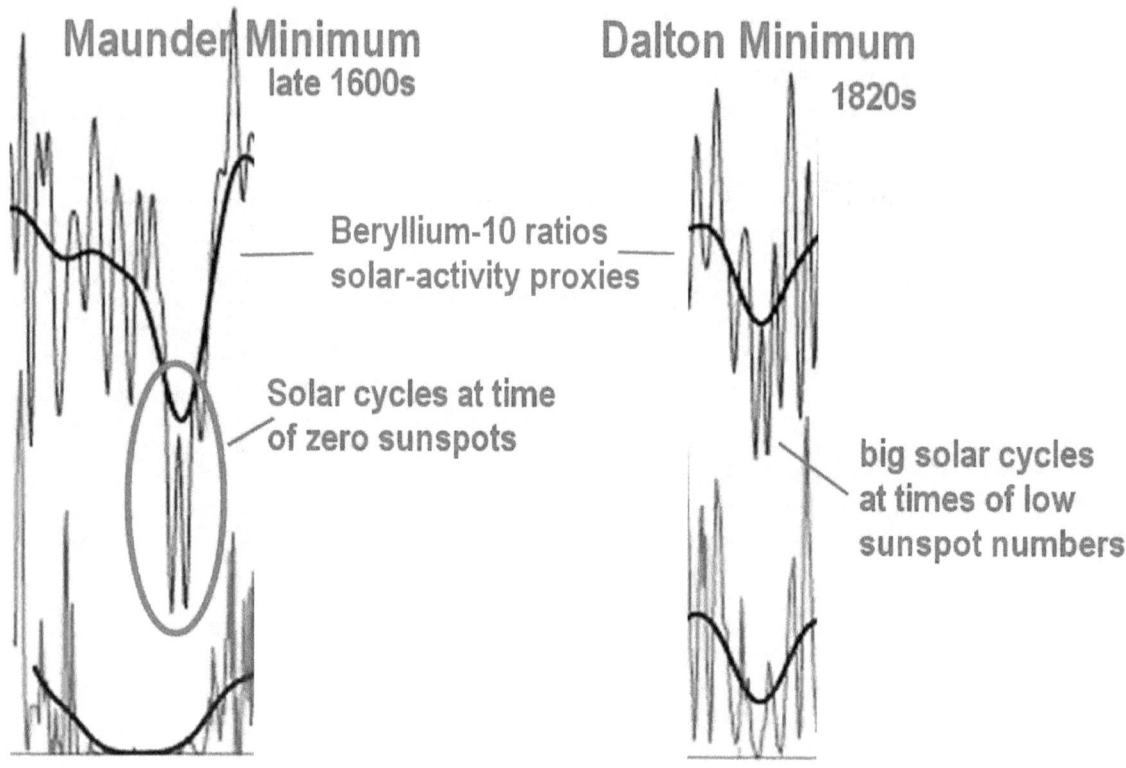

No, the Sun didn't go to sleep during the Maunder Minimum. Even during the times of zero sunspots, the solar cycles had continued, although at such a low level that they no longer caused sunspots on the Sun, however, they did cause corresponding changes in solar cosmic-ray flux. The cosmic-ray flux is measurable in ratios of the Beryllium-10 isotope that cosmic-ray effects generate in the atmosphere.

The solar cycles shown in blue are measurable proxies for solar activity, measured in Beryllium-10 ratios. Beryllium-10 is produced by cosmic-ray collisions with Nitrogen atoms and Oxygen atoms in the atmosphere. The isotope thereby becomes a measurable proxy for solar activity.

When the Sun is weak

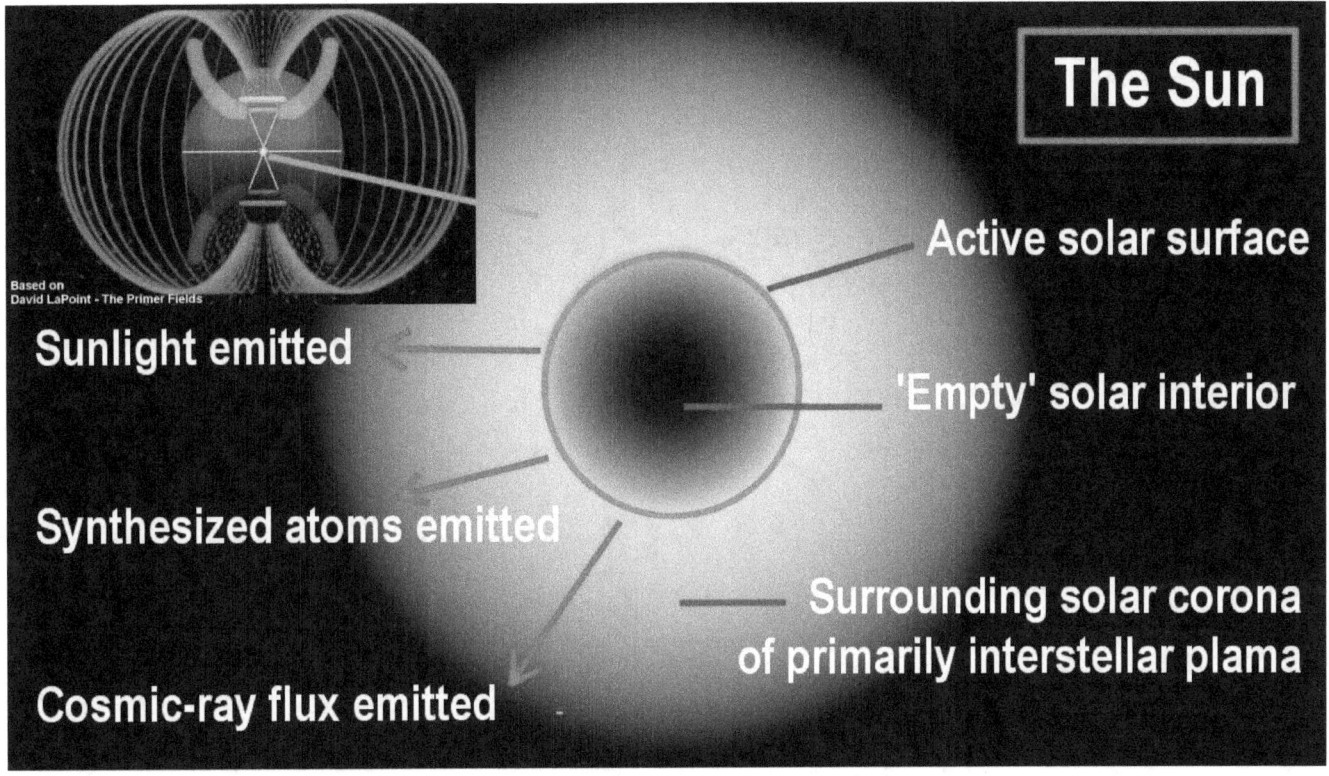

When the Sun is weak, the plasma shell around it is weak also, which enables larger volumes of cosmic rays to penetrate it and affect the Earth, and generate Beryllium-10 in the Earth's atmosphere.

Activity fluctuations occurred below the visible-threshold

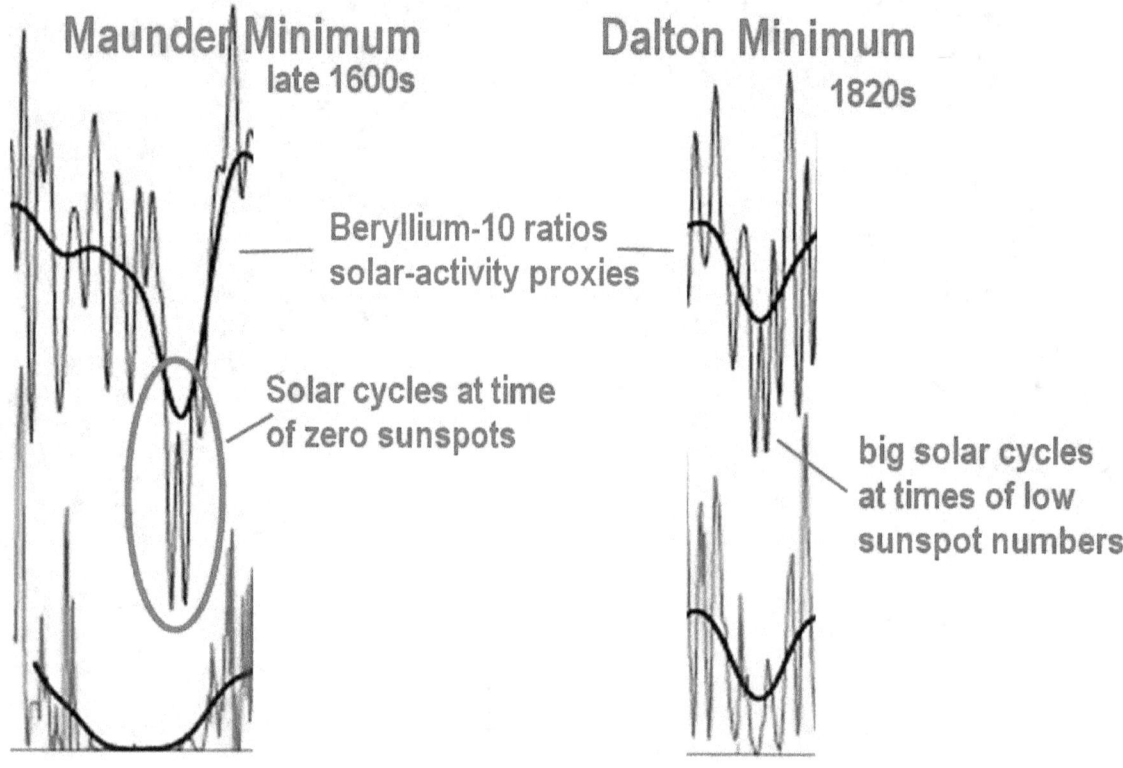

The Beryllium ratios, which are high when the Sun is weak, are presented inversely here, to illustrate the weakness of the Sun. The Sun was so weak during the Maunder Minimum that the activity fluctuations occurred below the visible-threshold.

A portion unfolds at the sub-visible level

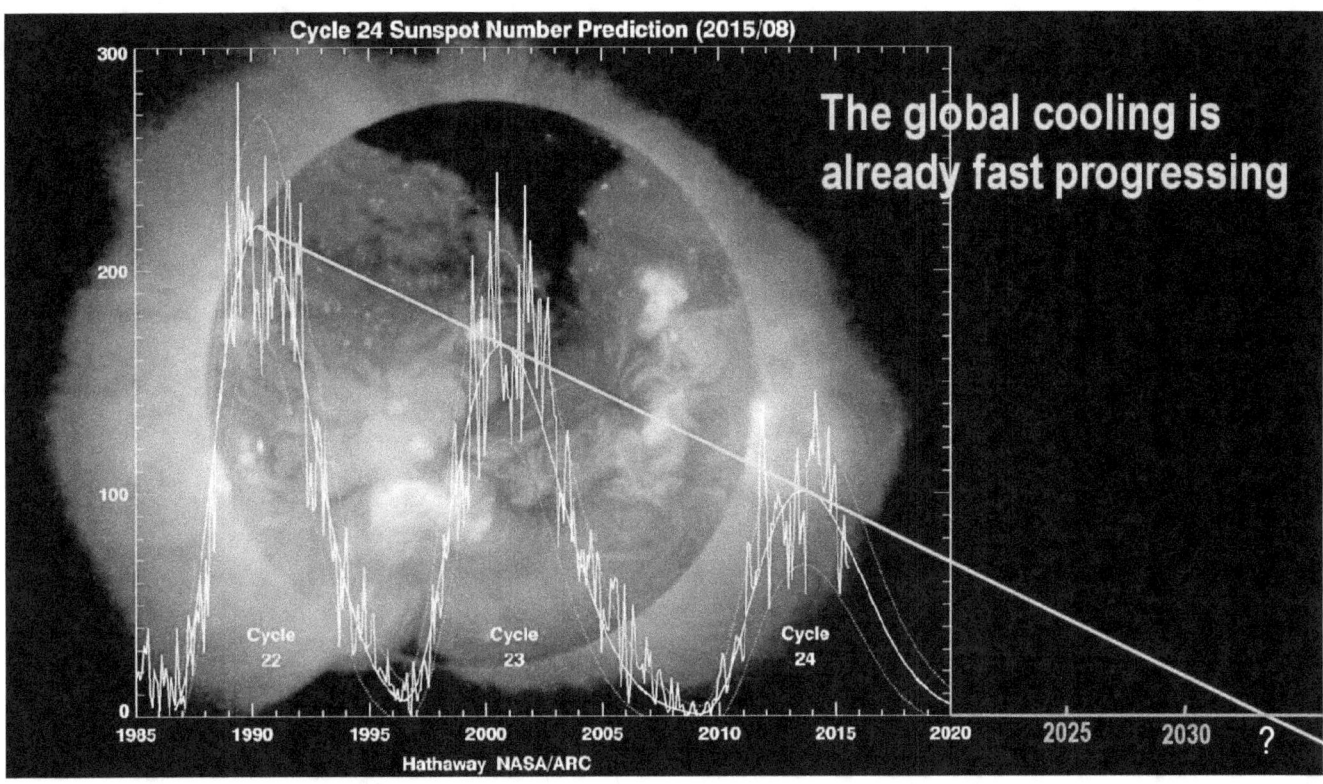

We see this type of process happening again in modern time.

When we see the solar cycles becoming smaller in terms of sunspot numbers, this means that a portion of the solar activity cycle unfolds at the sub-visible level, below the Zero-Sunspot level. We get into these situations typically at the beginning and end phases of a solar cycle.

They unfold at progressively lower levels

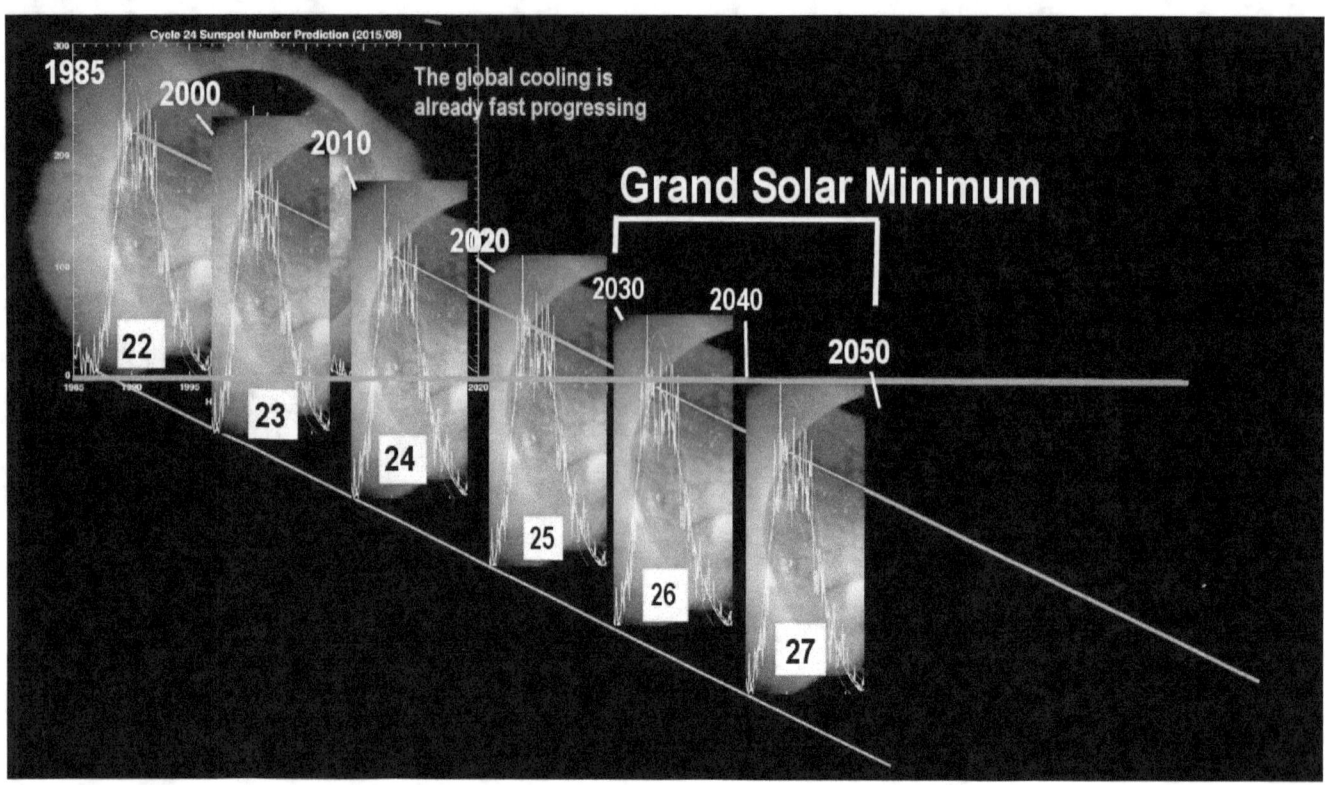

This means that the solar cycles themselves don't get smaller in amplitude. It only means that they unfold at progressively lower levels.

This is why the peak of cycle 24 is so low

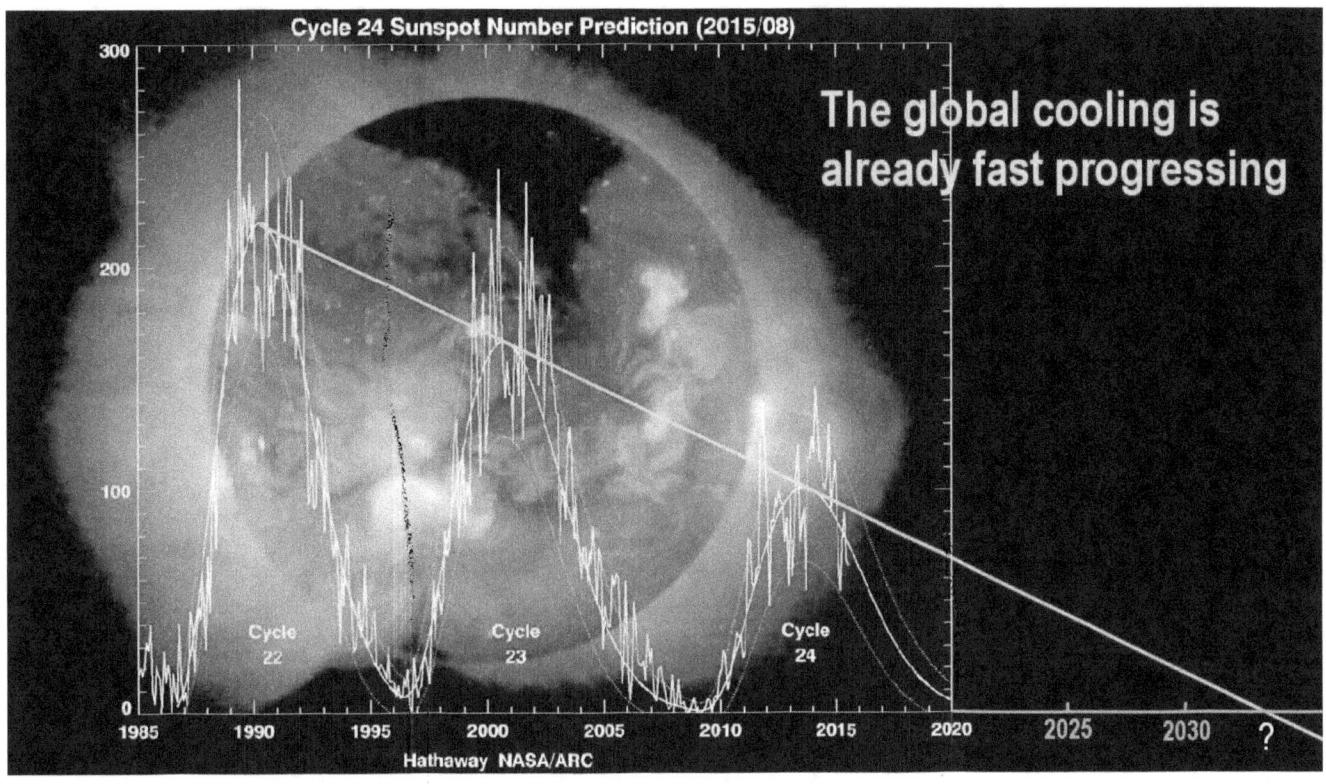

This is why the peak of cycle 24 is so low in sunspot numbers, as half of the cycle unfolds at the sub-visible level.

As we look ahead into the near future

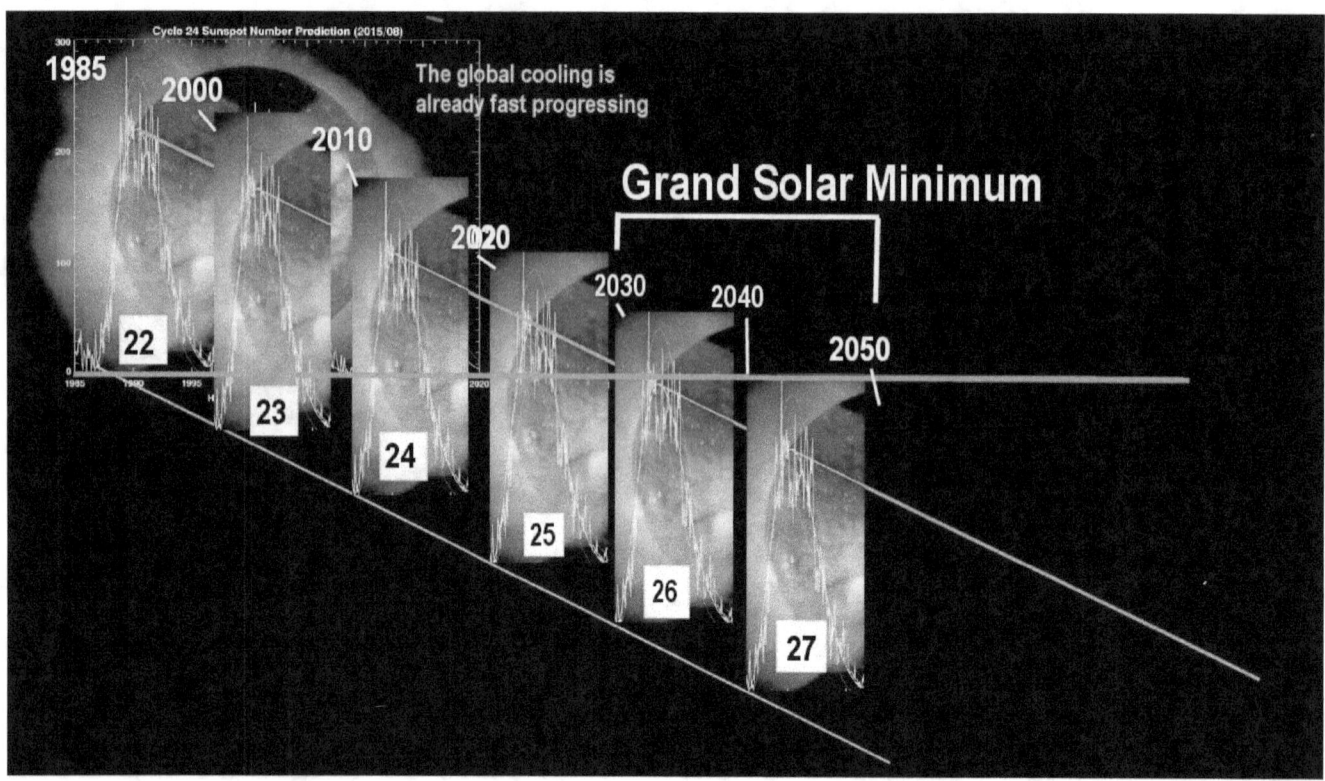

The correlation is significant in principle as we look ahead into the near future. The projection forward tells us that cycle 25 will become visible late into the cycle, be extremely weak in numbers, for a short period, if we see any sunspots at all.

Cycle 26 and 27 will be so weak that all of them unfold in the sub-visible, with zero sunspots on the face of the Sun, without a recovery happening this time. This projection tells us that cycle 27 will be 135% weaker than cycle 22 had been, in the 1990s.

A 135% collapse of the solar activity in the space of only 5 solar cycles, spanning less than 60 years, is huge. It is almost unpredictable in consequences. There likely won't be a recovery happening.

Grand solar Minimum in the 1970-80s

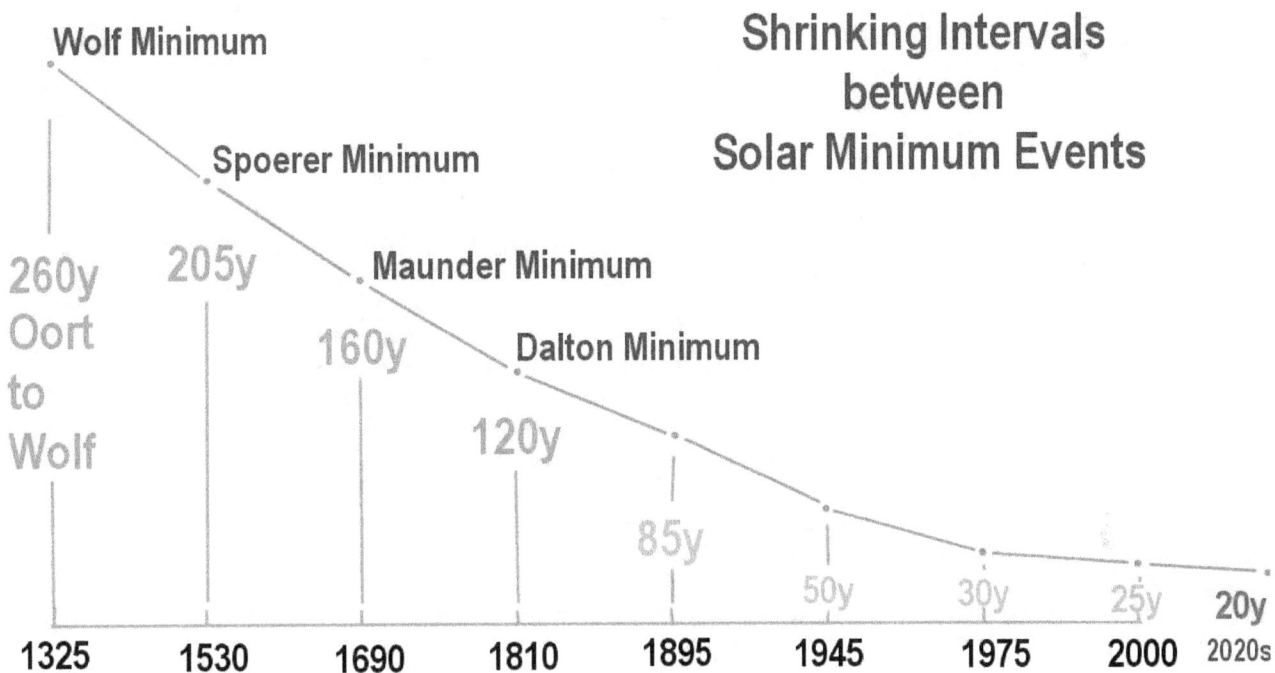

The last Grand solar Minimum event that we see in the sunspot numbers had occurred in the 1970-80s timeframe. We saw a recovery thereafter of 50 sunspots.

If such a 'big' recovery should happen again after the next low point projected for 2017, which falls into the current solar solar cycle 24, we might see increasing sunspot numbers during solar cycle 25, which would reflect the Sun recovering from the minimal period in cycle 24.

Minimal periods getting smaller

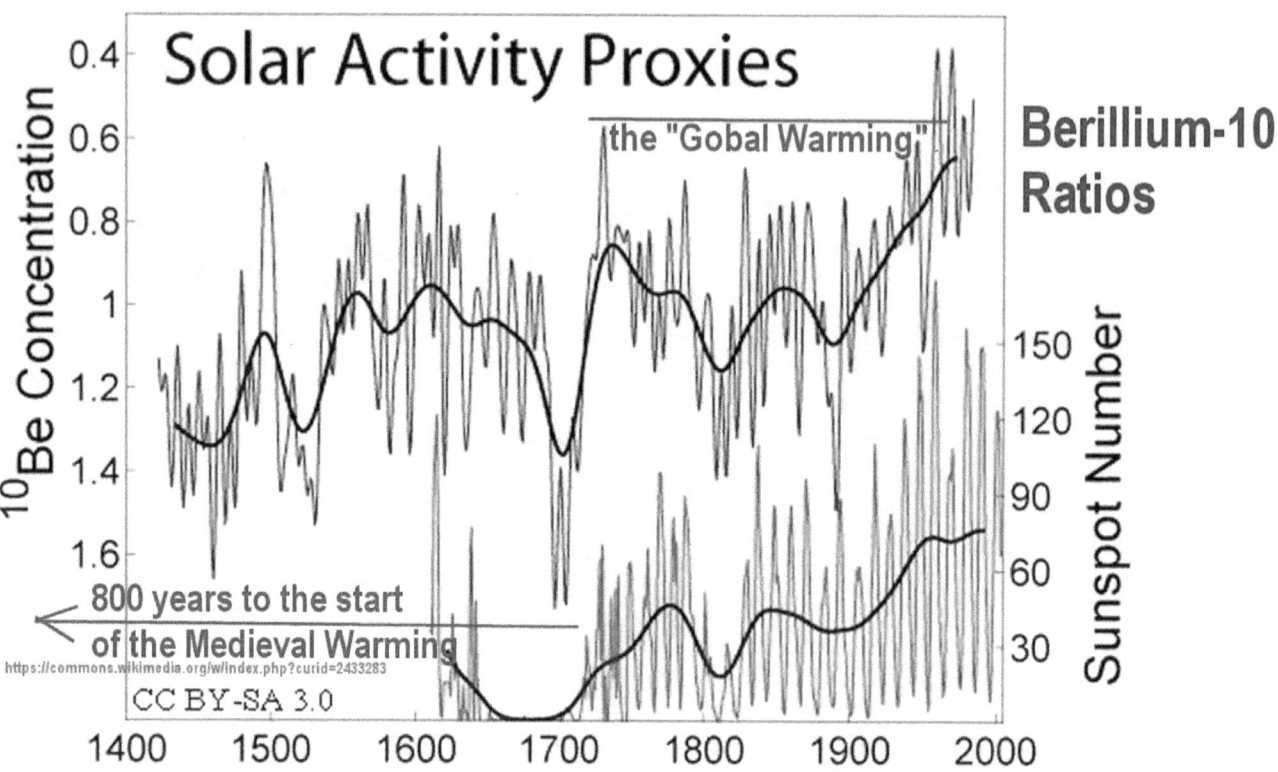

However, the amplitudes of minimal periods have been getting progressively smaller, in line with the shorter intervals. The last large minimum cycle was the Dalton Minimum of 1810. It caused a noticeable recovery in the sunspot numbers However, the next such cycle in the 1890s was much weaker. It was so feebly expressed that it isn't regarded as minimum event anymore. The same is apparent about the minimum cycles that followed, including that for cycle 24.

The Dalton Minimum of 1810

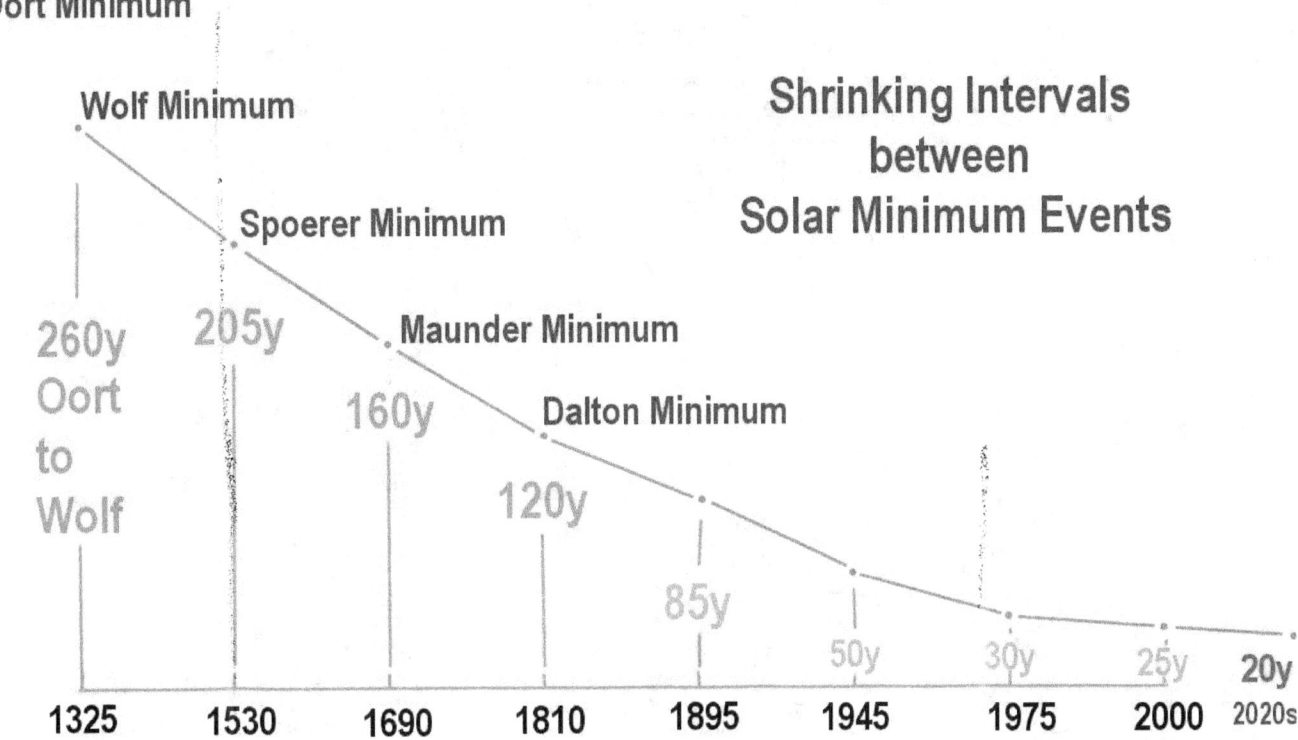

The Dalton Minimum of 1810 had occurred at a time when the solar system was still relatively strong and cycle intervals long. Now that the interval has shrunk to less than a third, the amplitude has shrunk with it accordingly, to a level that is no longer of any significance.

The expected Grand Solar Minimum of the 2030s and 40s

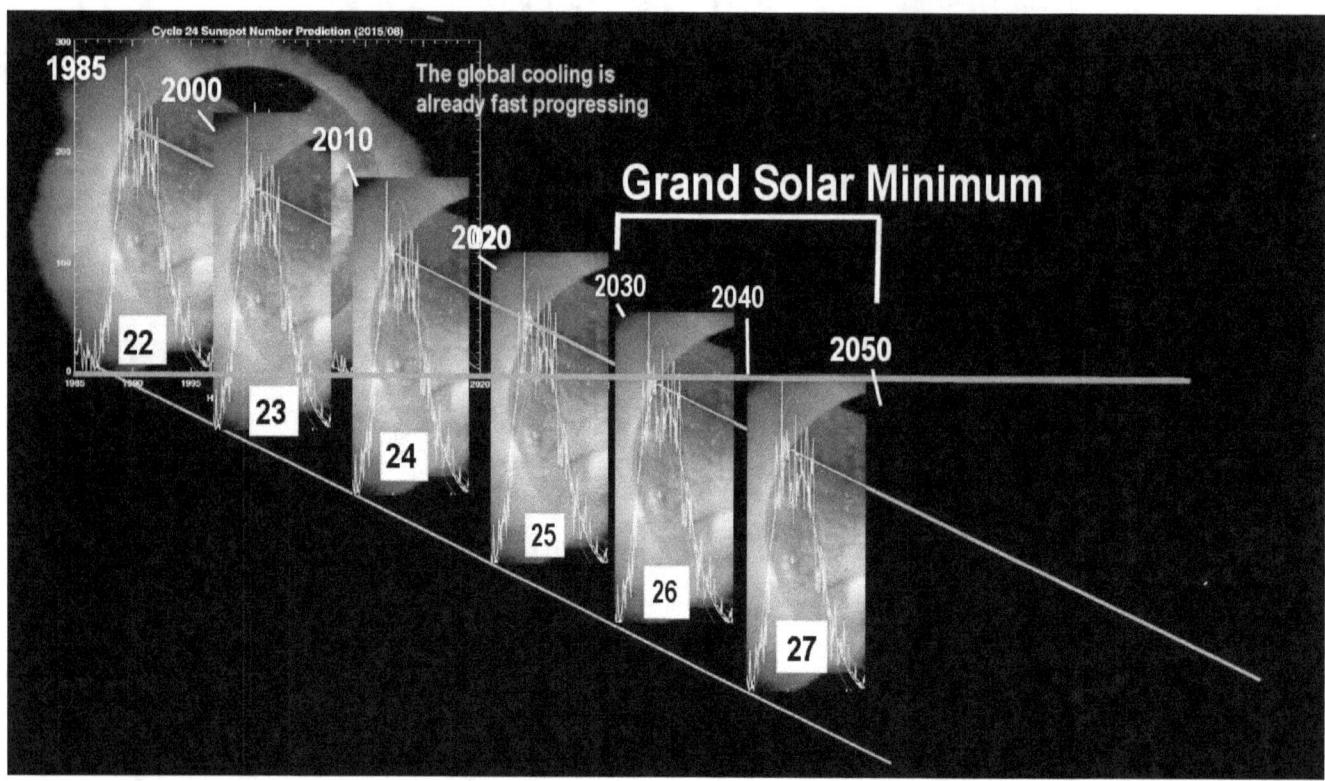

This means that we won't likely see a noticeable interruption of the sharply diminishing solar cycles that began with the year 2000. This also means that the expected Grand Solar Minimum of the 2030s and 40s, is not a cyclical repetition of the historic Maunder Minimum of the 1600s in the Little Ice Age period, but has a different cause.

The long down-trend that we see here

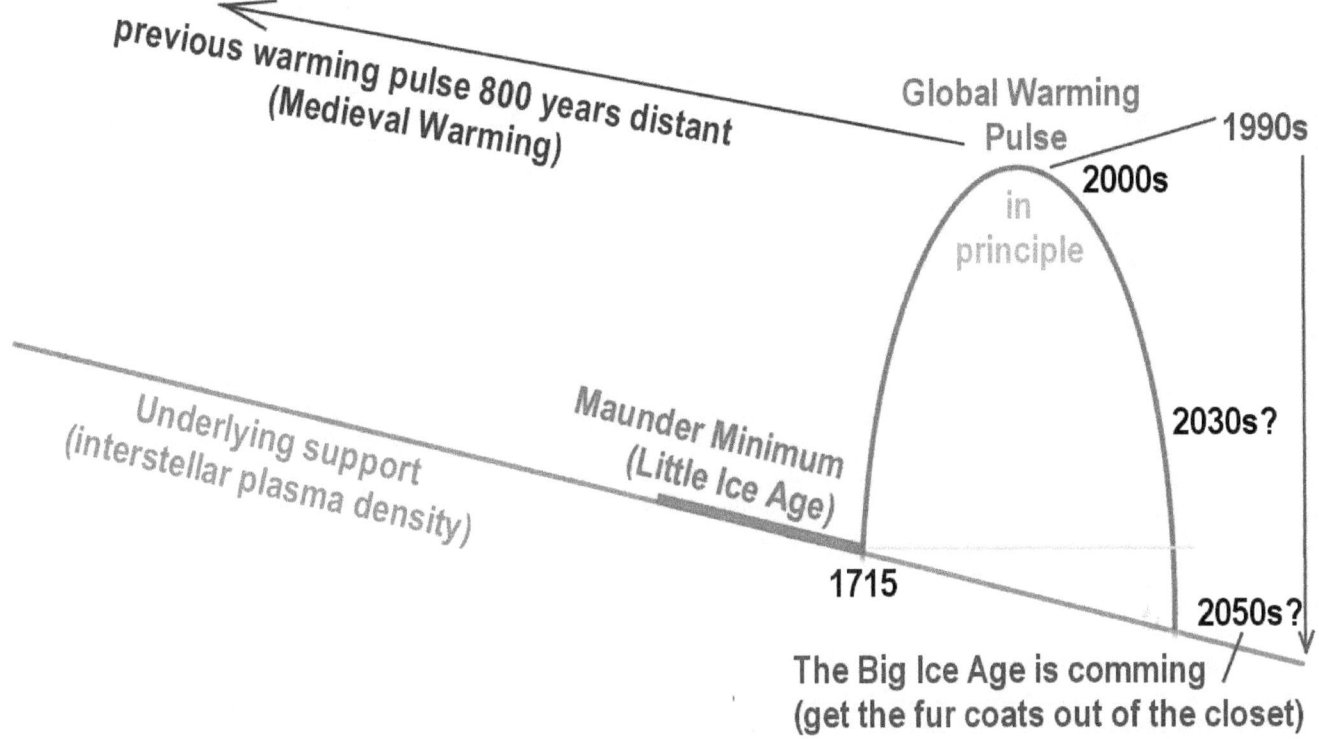

The long down-trend that we see here, that began with solar cycle 22 in the 1990s is evidently the result of the global warming pulse now ending that had lifted us out of the Little Ice Age in 1715 and gave us the global warming period that followed. That's what is now ending.

The historic solar minimum cycles are so weak

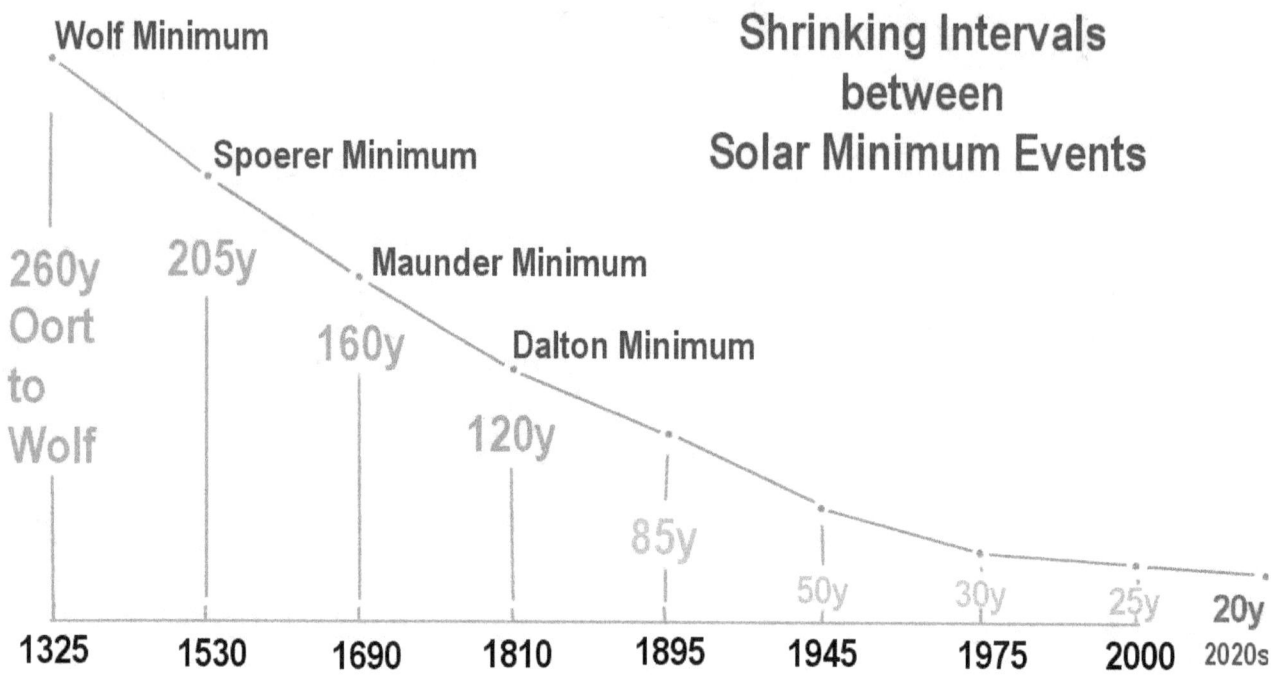

The historic solar minimum cycles are so weak that they are, for all practical considerations, history. They lack the intensity to affect us anymore.

The weakening of the Sun is symptomatic

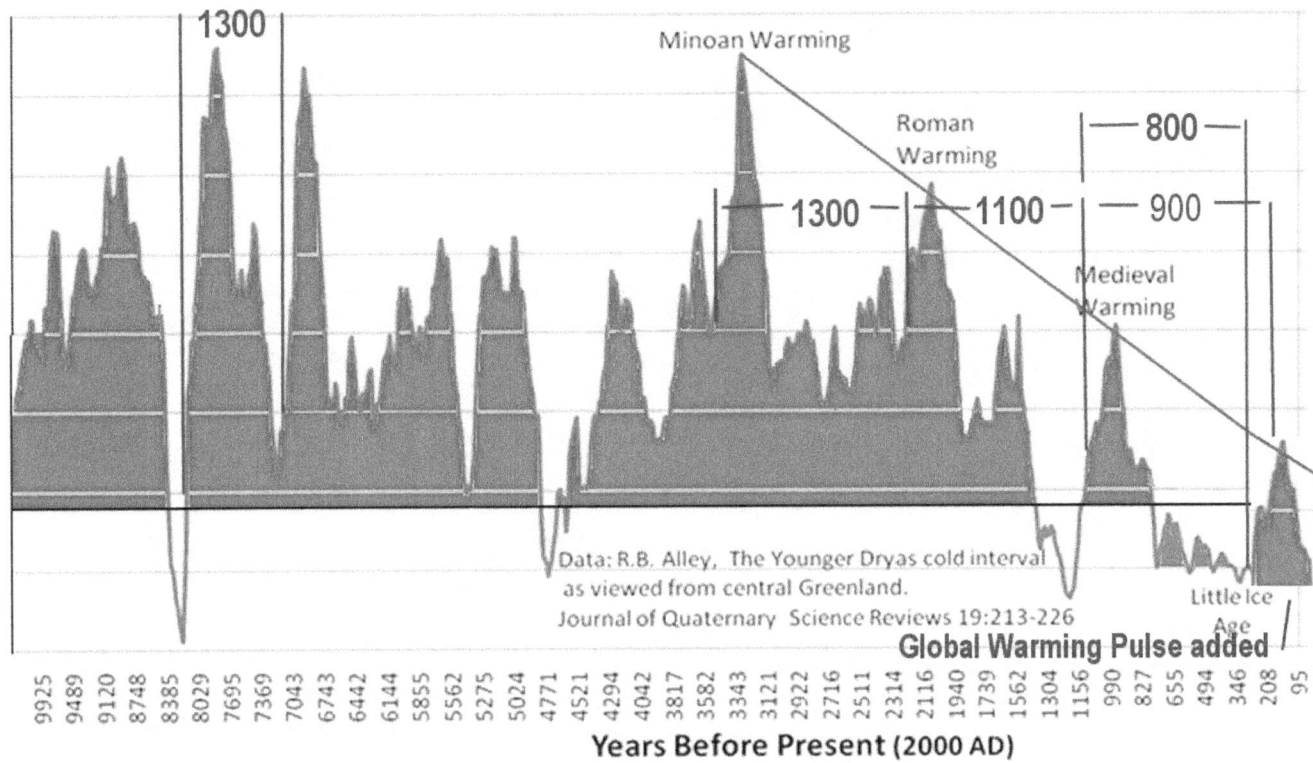

The weakening of the Sun and its solar activity will likely continue and may even increase dramatically, which is symptomatic of the sharp collapse that we have seen in the past for the historic warming spikes.

The underlying level of the interstellar plasma density

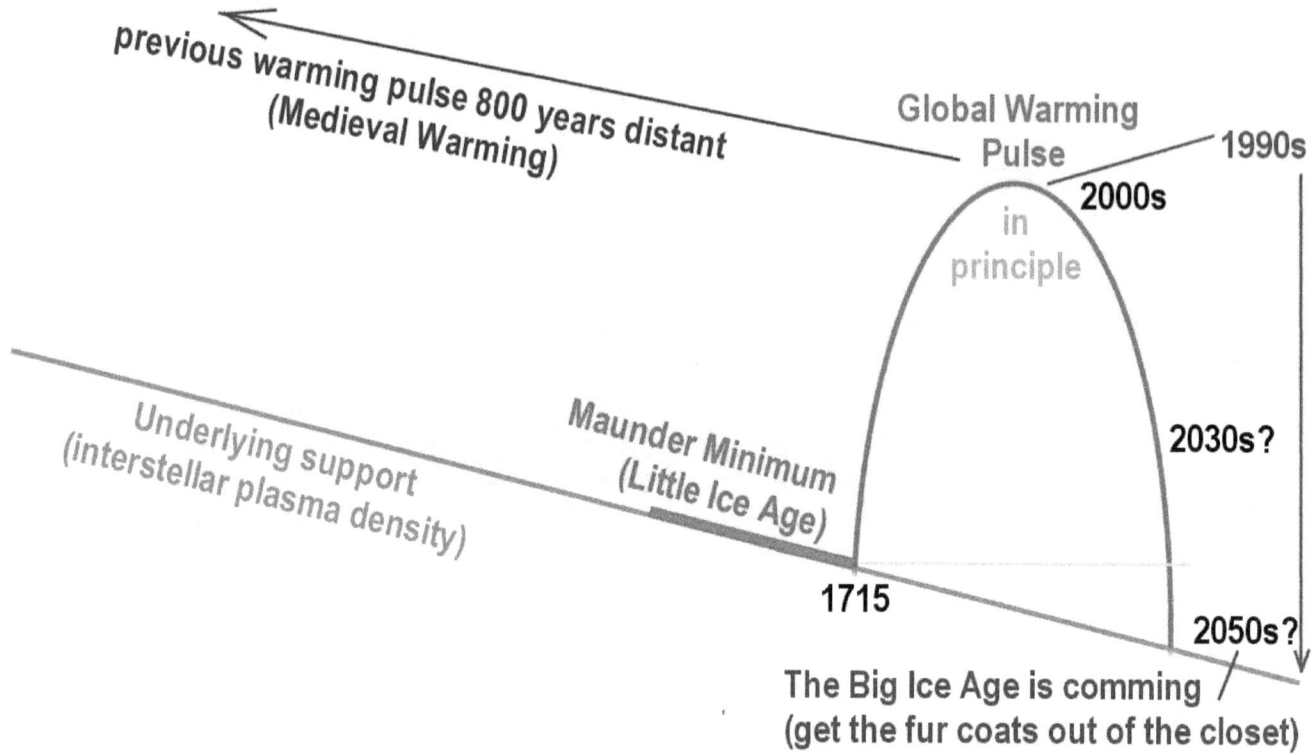

Nor will the 'collapse' of the current warming spike stop at the level that it started with, because the underlying level of the interstellar plasma density, has over the years continued to diminish further in the background since the time of the Maunder Minimum, possibly even at an increasing rate of diminishment.

At some point on the steep slope

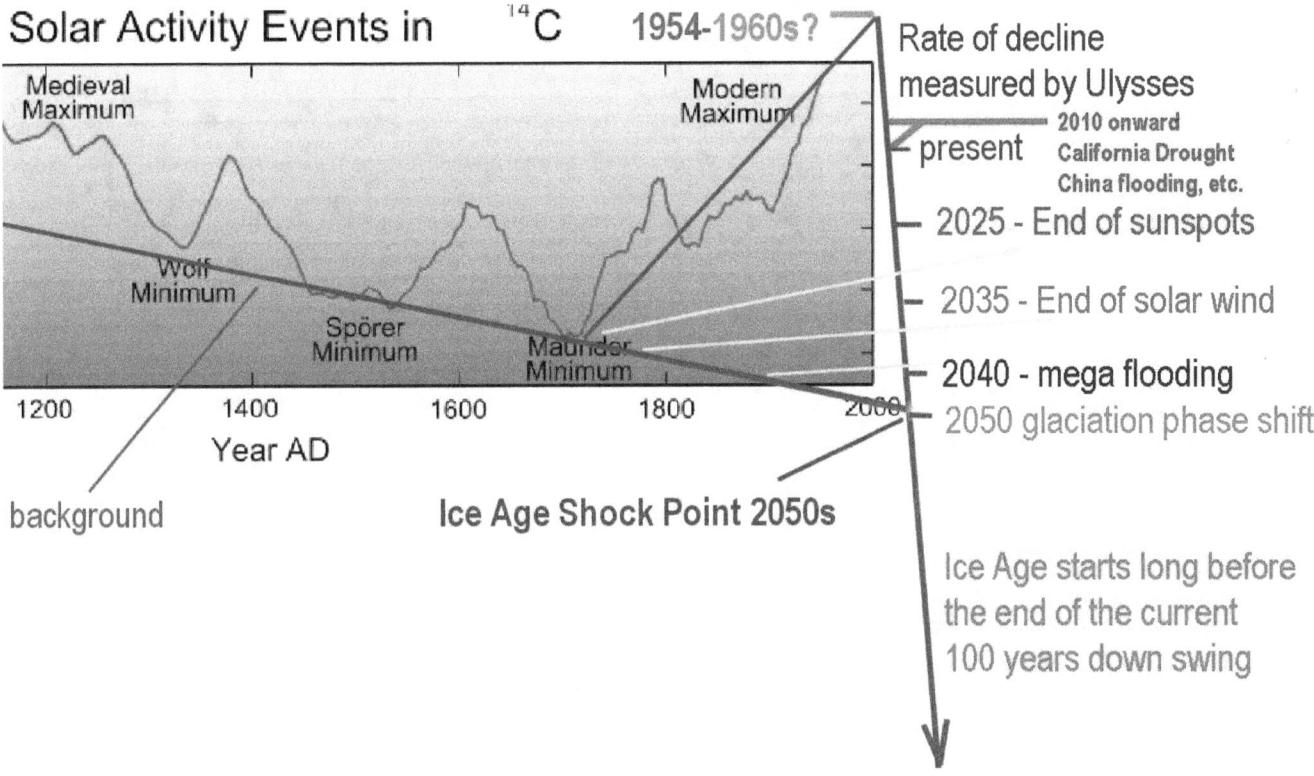

At some point on the steep slope of the collapsing Global Warming pulse, the critical threshold will be crossed where the primer fields collapse and the next Ice Age glaciation will begin.

We don't know what the minimal threshold level is

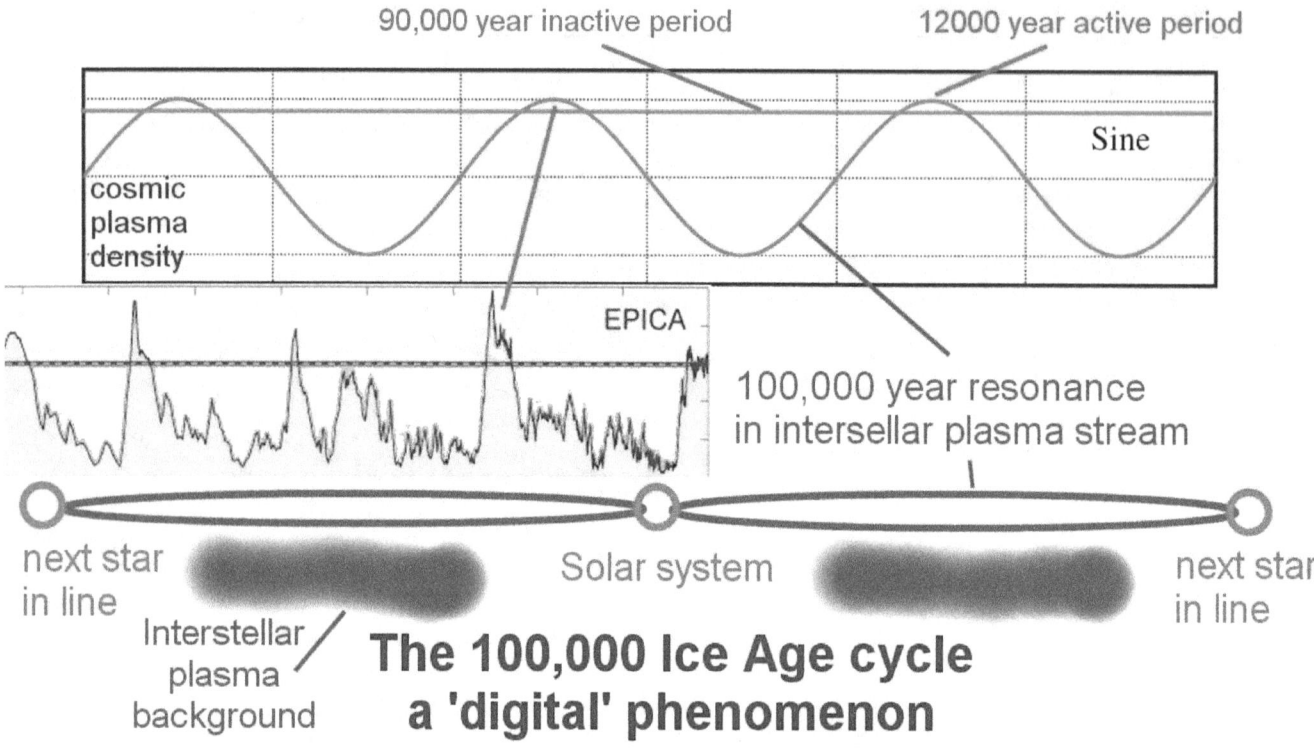

The potential timing of the primer fields collapsing is shrouded with great uncertainty, because we don't know what the minimal threshold level is beyond which the primer fields collapse.

We can only say with a high degree of certainty

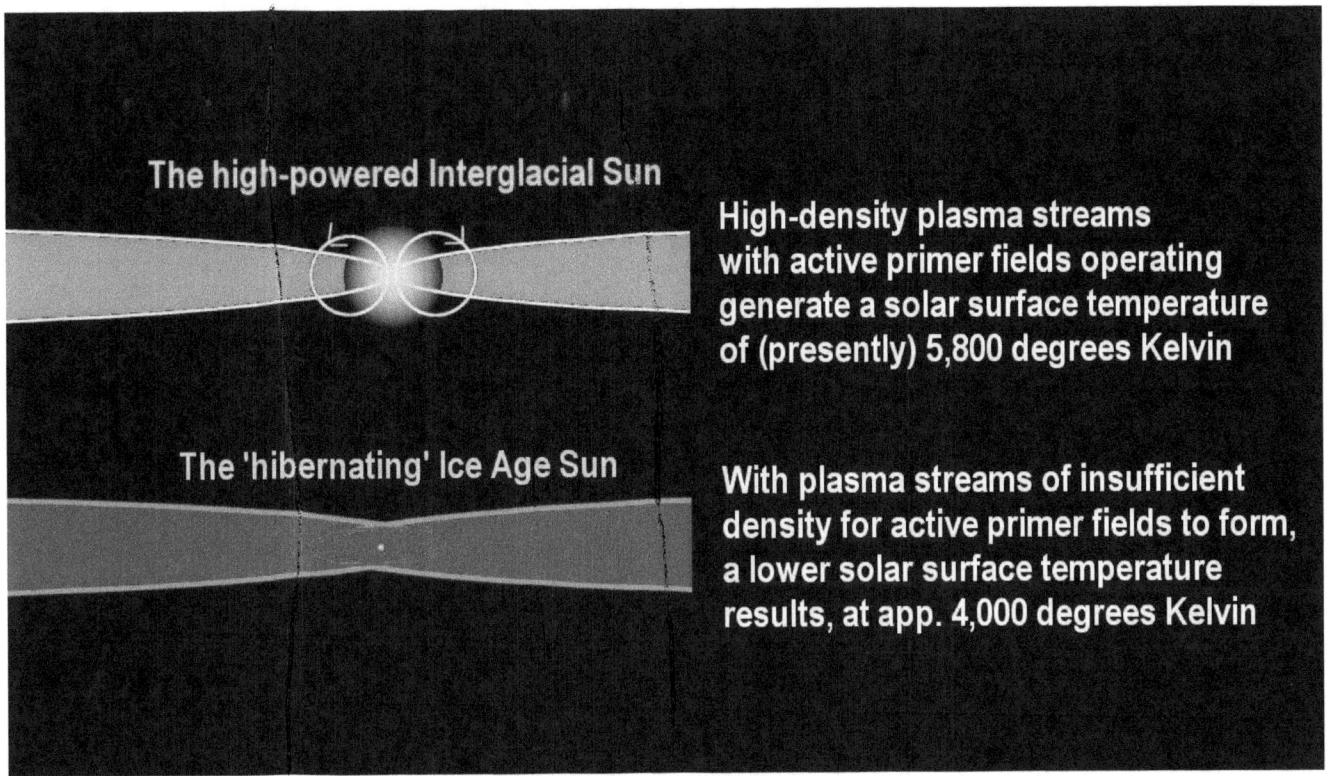

We can only say with a high degree of certainty, that when this threshold is crossed at which the primer fields fail, the Sun will fall back to its long 90,000-year hibernation of the next period of glaciation.

With so many aspects of the solar system now all rapidly diminishing, the best prediction that one can offer at this time, is that the phase shift will happen 'soon,' that it is 'near,' whatever this may mean.

As it has been noted by Jaworowski

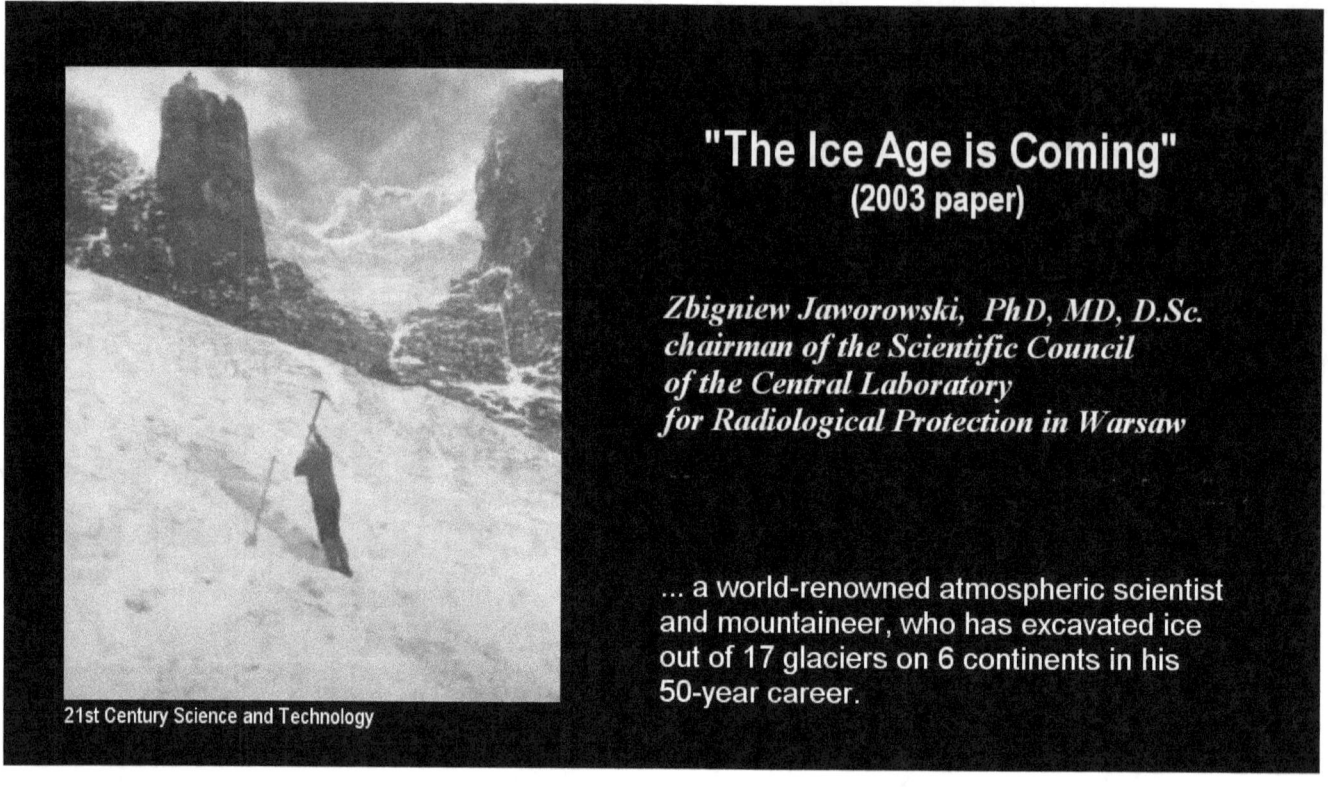

As it has been noted by Jaworowski in 2003, the "transition" will likely happen without warning.

Get the fur coats out

Get the fur coats out. The warning bells are already ringing.

➤ The 'sinking' of the 'Rock of Gibraltar'

> The 'sinking' of the 'Rock of Gibraltar'
> **The final event**

The 'sinking' of the 'Rock of Gibraltar' - The final event.

The rate of the collapse of 30% in 10 years

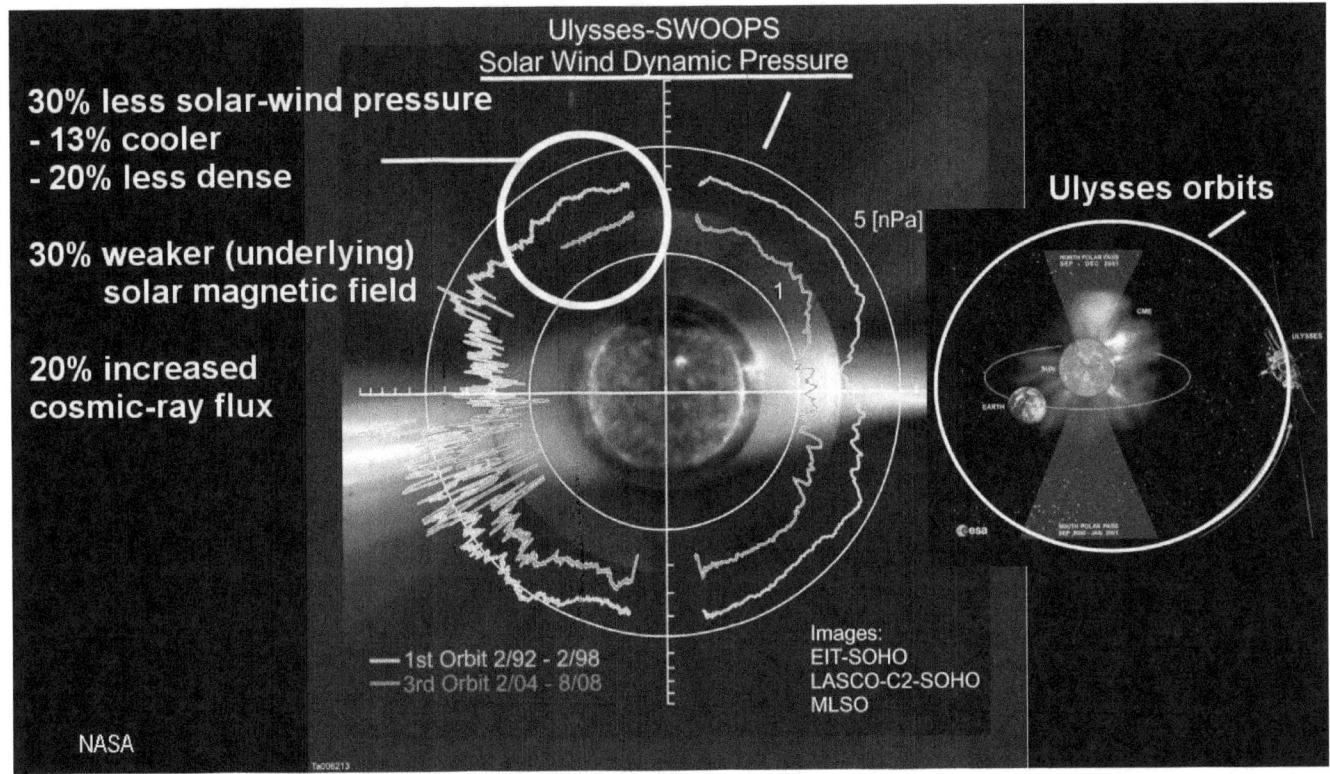

The rate of the collapse of the solar-wind pressure, of 30% in 10 years, that the Ulysses spacecraft has measured, gives us an indication of how close we have come to the phase shift event.

The rate of diminishment of the solar-wind pressure

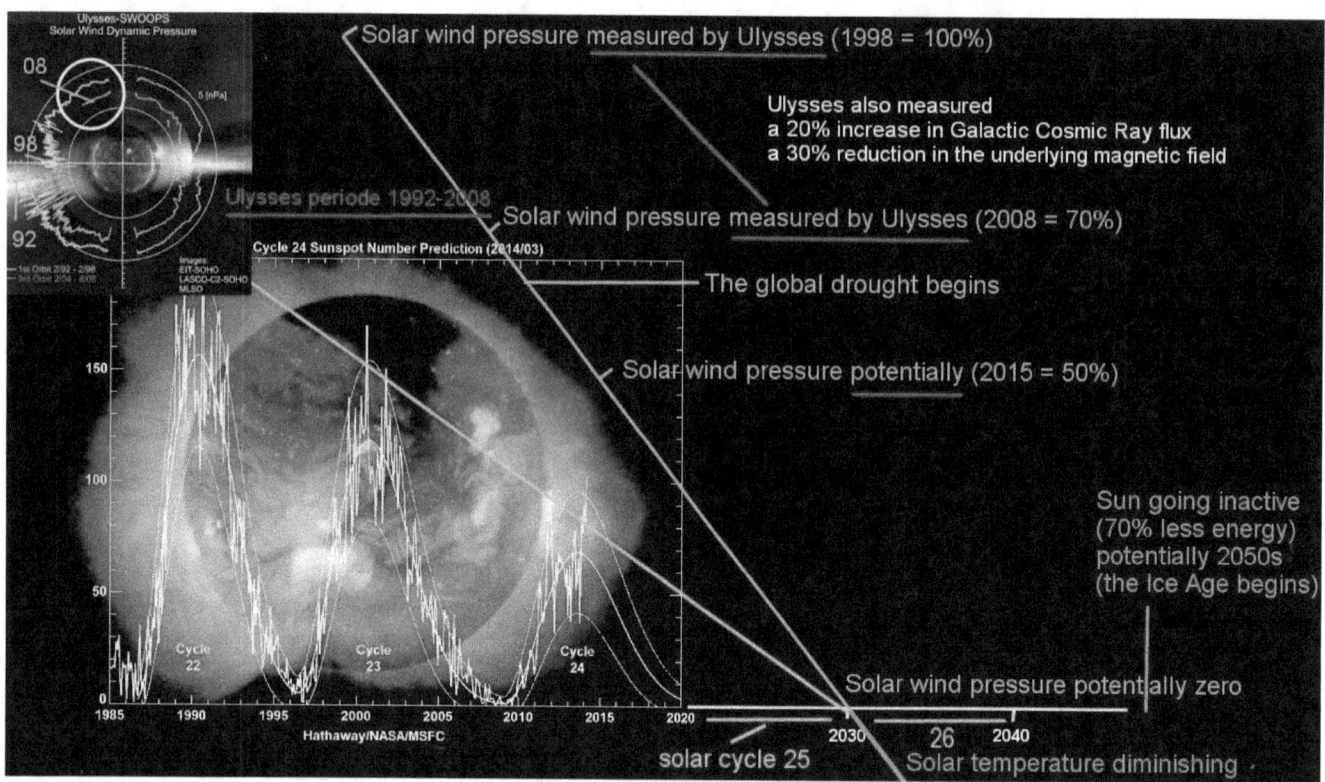

The rate of diminishment of the solar-wind pressure closely matches the rate of diminishment of the solar activity cycles. The two are the same. This means that the diminishing solar activity intensity is the driving cause for the diminishing solar-wind intensity.

The solar-wind pressure cannot diminish below zero

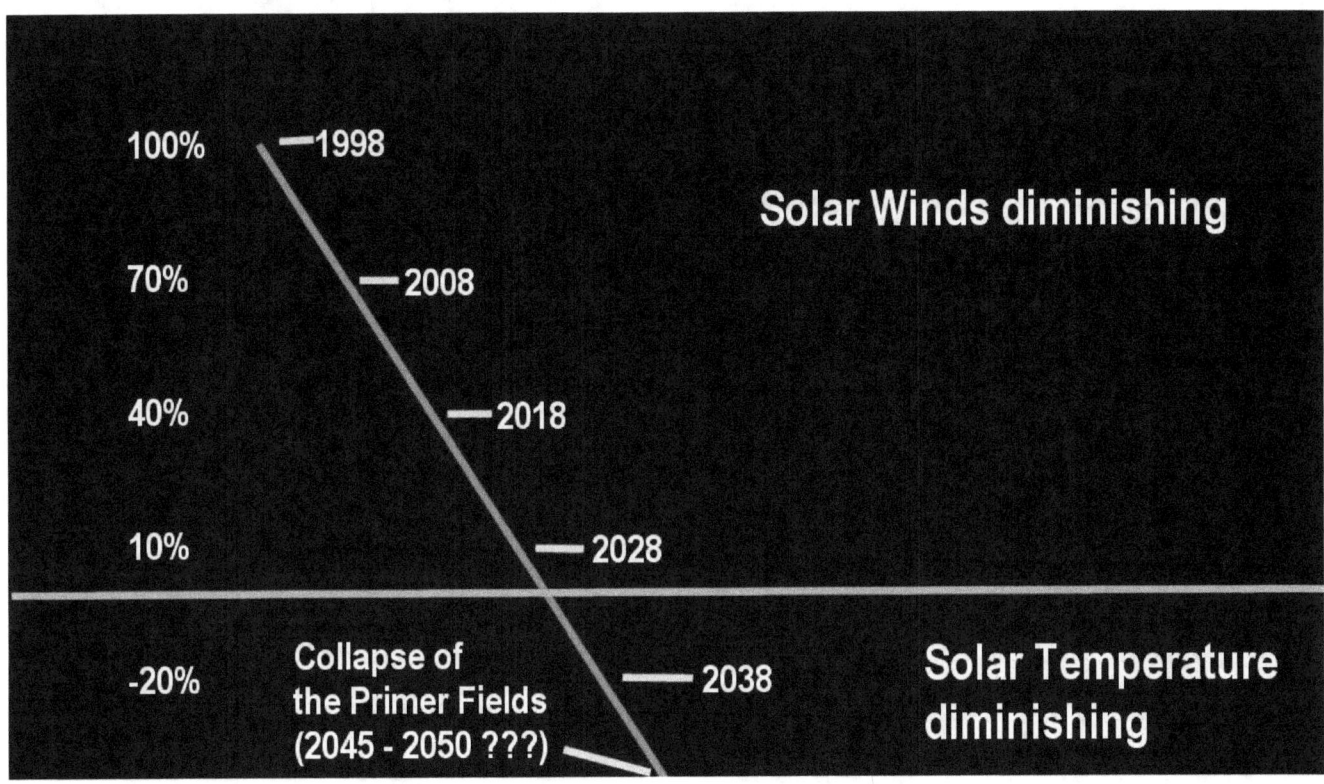

However, the solar-wind pressure has its own built-in threshold. It cannot diminish below zero, while the solar-activity intensity continues to diminish.

A new phase in the solar collapse

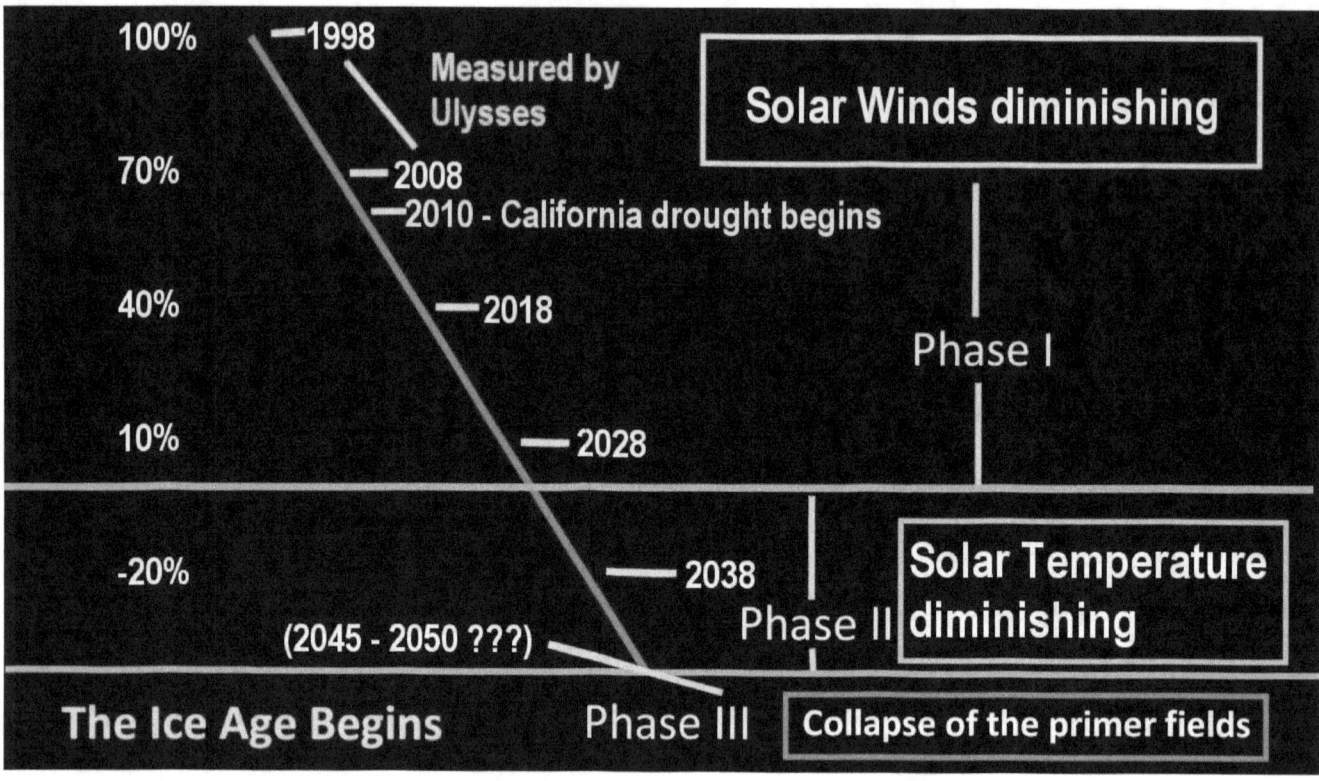

Here a new phase in the solar collapse process begins. The new phase reflects itself in the Sun's surface temperature beginning to diminish.

The Sun's surface temperature on course to its 'sinking'

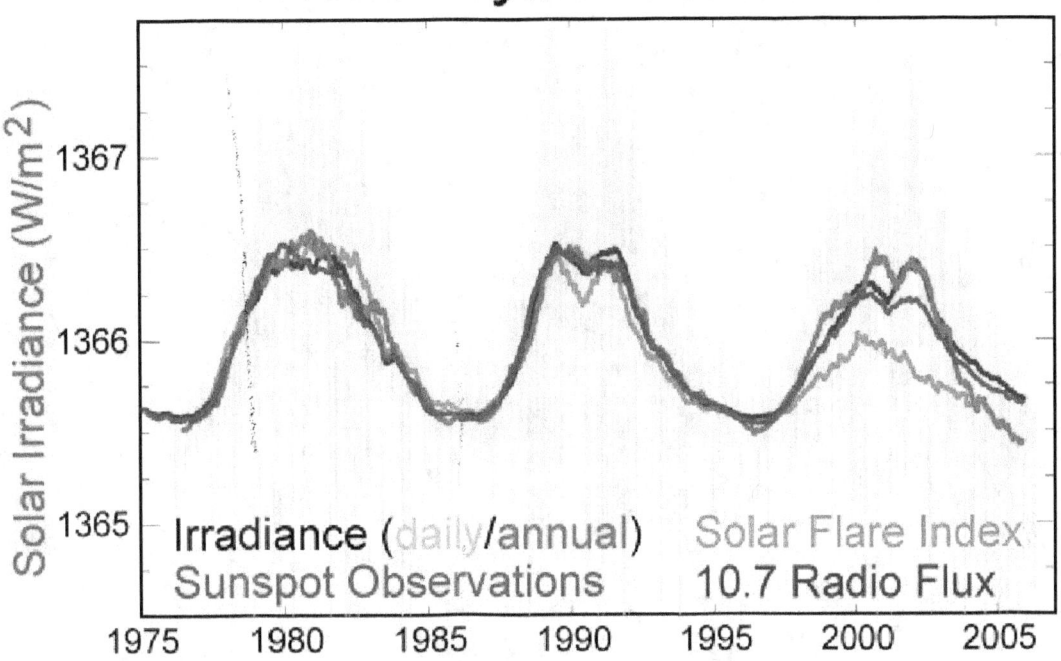

The 'Rock of Gibraltar' - the solar system's most unchanging feature - the Sun's surface temperature - its radiation intensity - is on course to its 'sinking'.

Uncertainty, is too mild a term

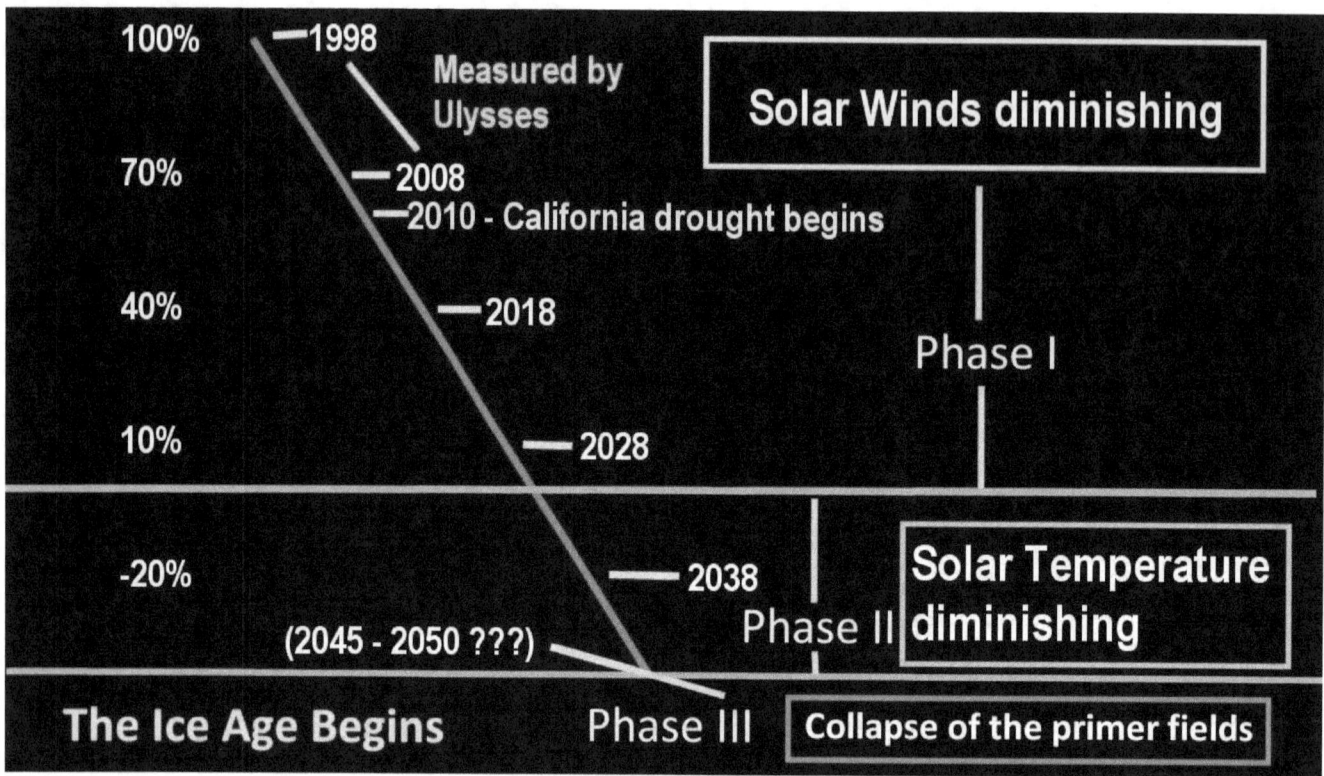

A new phase of uncertainty begins when the solar wind stops and the Sun looses steam and becomes colder. Uncertainty, is too mild a term, because of the lack of any reasonable yardstick for measuring the effect of the new phase in solar activity diminishment.

Until this point in the collapse process, the solar wind had vented off excess plasma pressure, which had kept the surface fusion action at a near constant level as we presently have it. When this regulating feature ends, which is poised to happen in the 2030s, we enter totally unpredictable territory.

This is bigger than the solar cycles slowing down. And it will be the final act in the diminishing of the solar system that began more than 2000 years ago. It will be the final act, because everything else is already on the path of diminishing.

The rate of collapse in solar activity

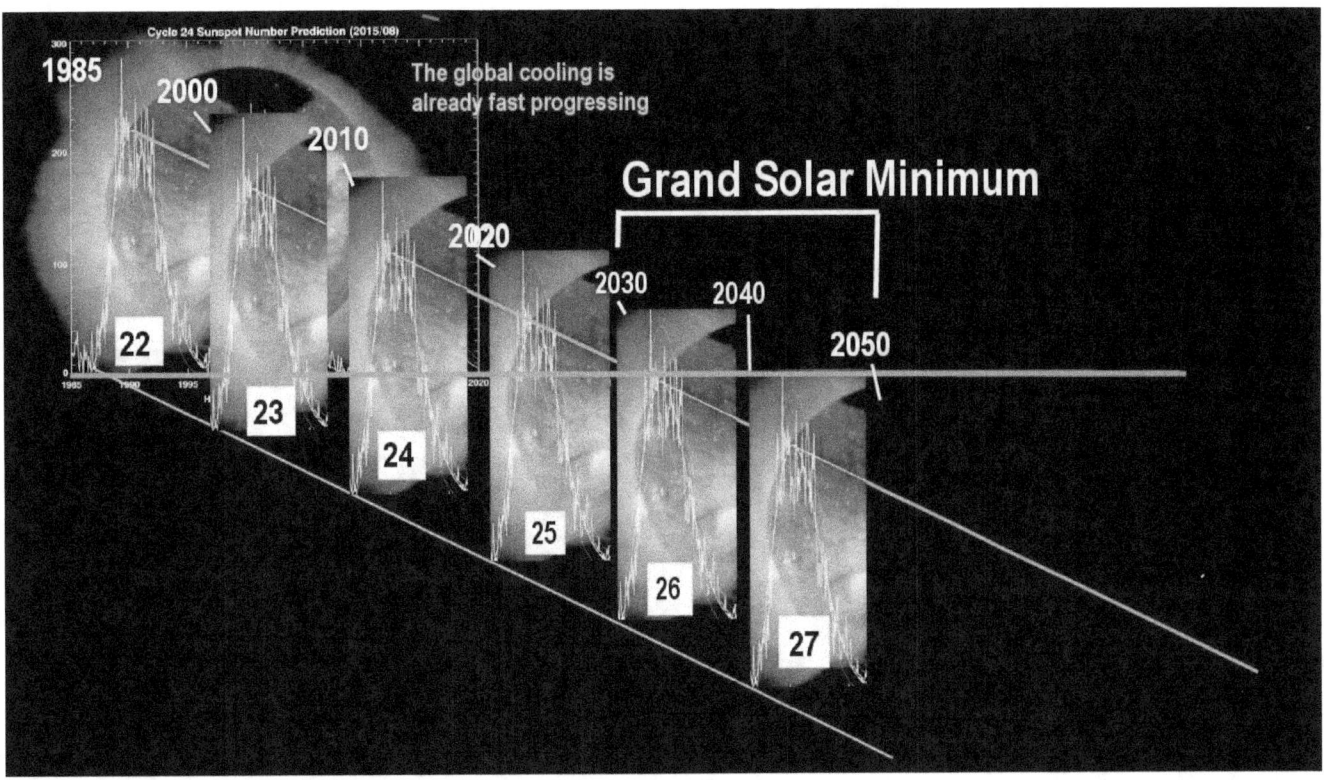

The rate of collapse that we see already reflected in solar activity cycles is real. It will continue.

Collapse measured in space by Ulysses

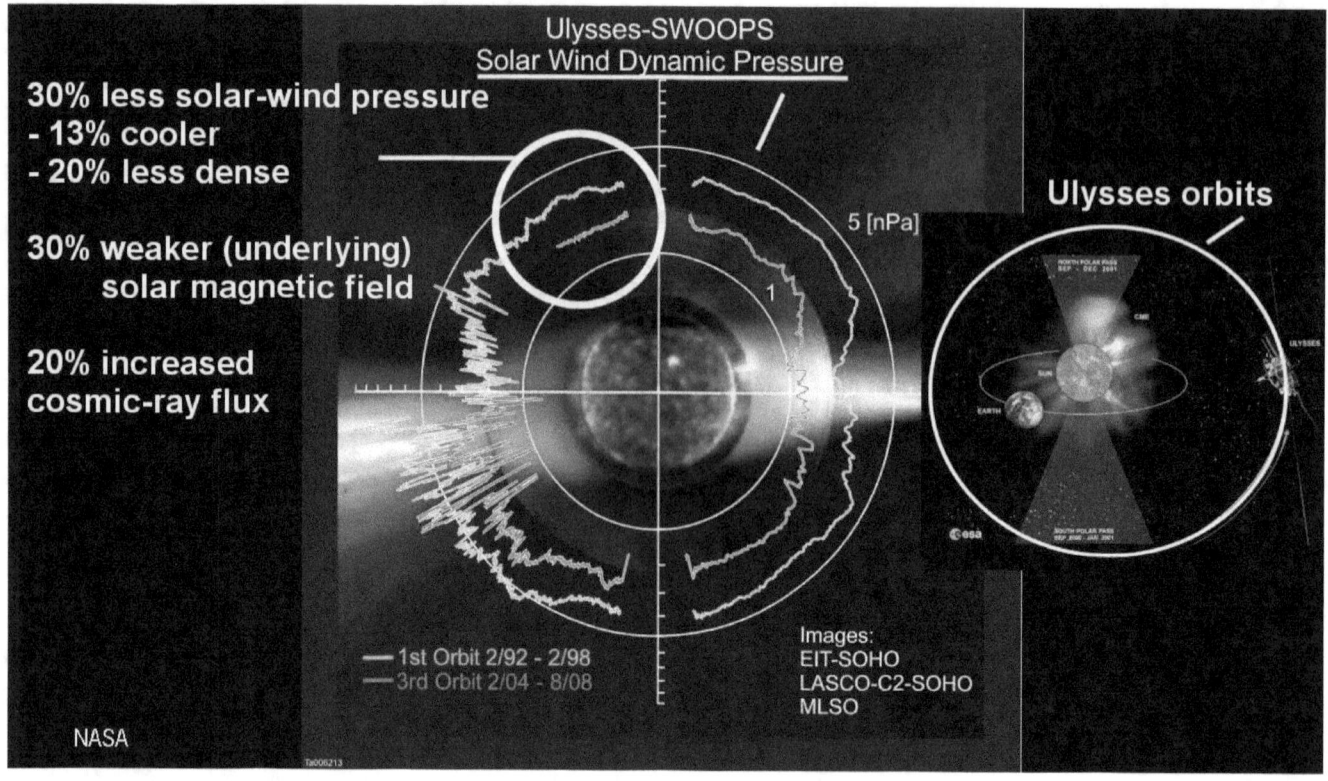

As I said, the same rate of collapse was measured in space by the Ulysses spacecraft, in terms of the collapsing solar-wind pressure, which the spacecraft saw only the beginning of.

Solar wind zero level in the 2030s

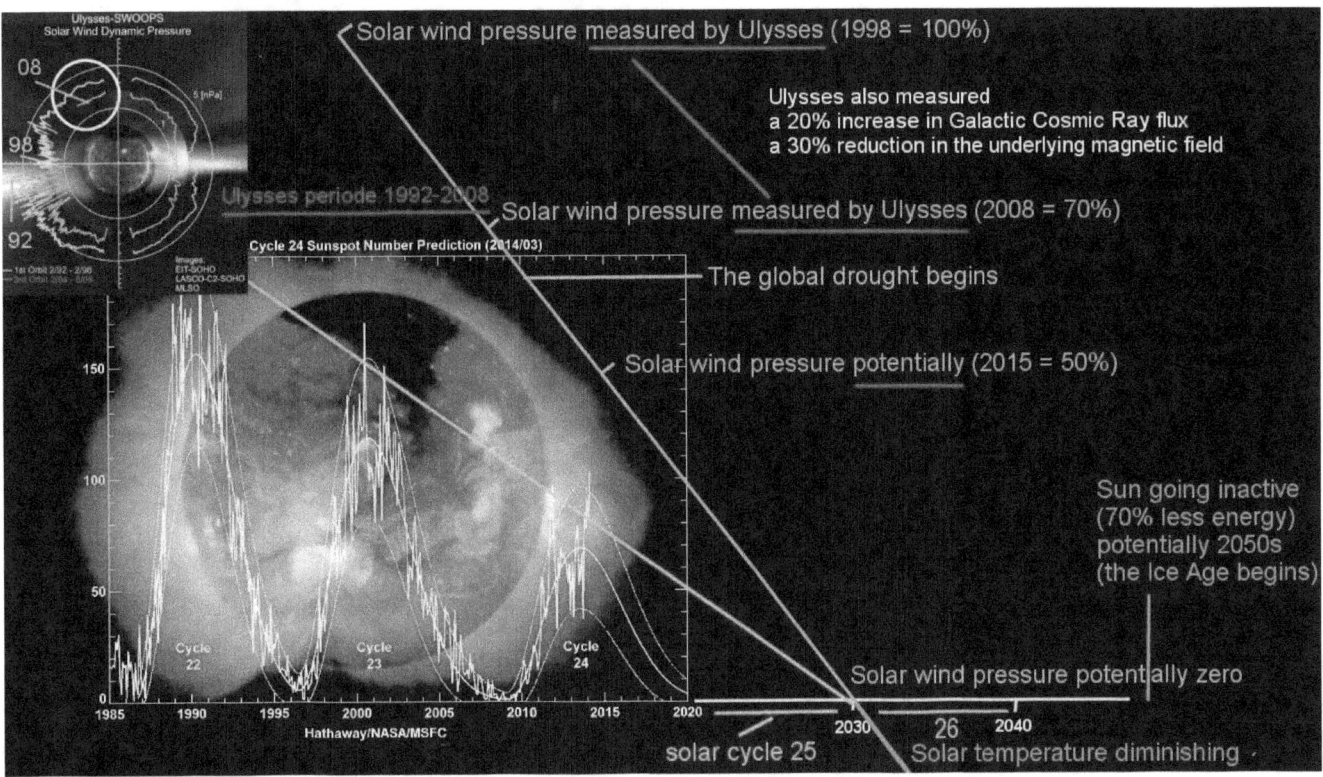

With the measured rate projected forward. the solar wind will reach the zero level in the 2030s. The zero-level is totally predictable.

Also, the timing of the new phase takes us directly in the timeframe of the Grand Solar Minimum of the 2030s.

Services the solar wind does provide

For this reason, let's explore what services the solar wind does provide, which we will no longer have, when it stops.

Solar wind to a kettle boiling off steam

Maximum temperature of liquid water at ambient pressure is 100 degrees Celsius: The Boiling Point

If one compares the solar wind to a kettle boiling off steam that has its heat turned down low, by which the steam stops and the water cools, a point will be reached on the Sun in a similar manner, by which its surface temperature becomes colder after the solar wind drops to zero. Obviously, there is more involved than just a change in temperature.

The cooling of the Sun, will begin in the 2030s

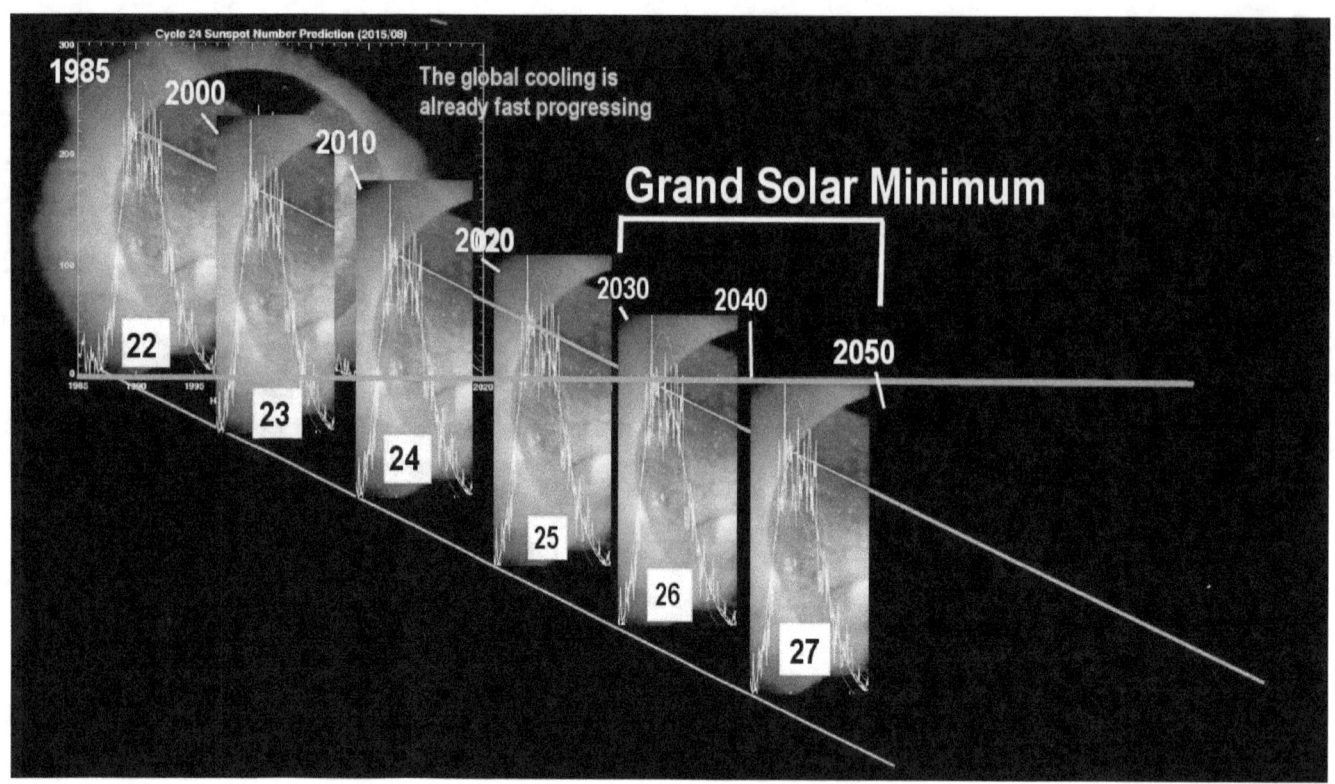

At the current rate of collapse of the solar wind, the cooling of the Sun, will begin in the 2030s, in the background of the Grand Solar Minimum.

The solar wind purging the plasma-fusion products

While the solar wind has no direct significant impact on the climate on Earth, its ending is significant as a change in the dynamic process that affects the operational principles of the Sun, where the solar wind plays a role.

The solar wind not only vents off excess plasma from the fusion cells on the surface of the Sun. It also aids the fusion cells in purging the plasma-fusion products.

The fusion products will clog up the fusion cells

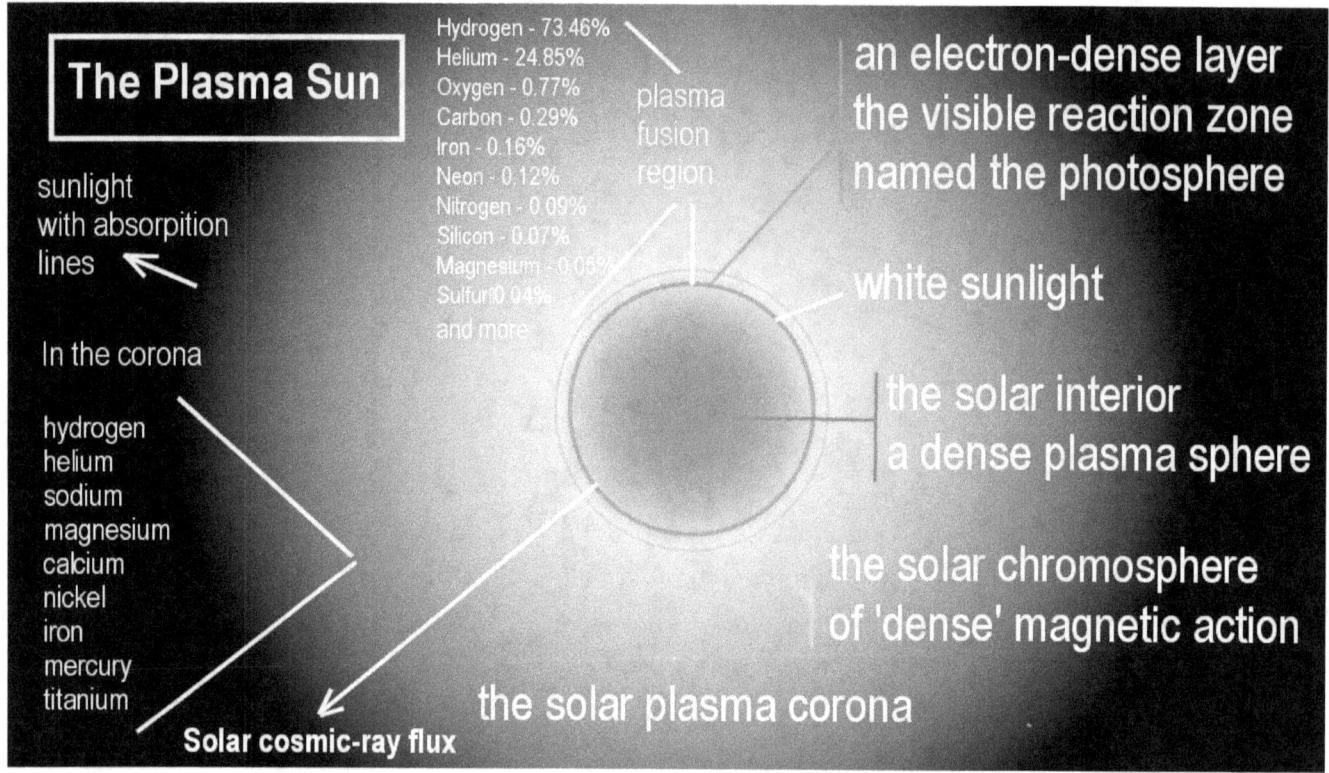

The fusion products are the atomic elements that the plasma-fusion creates. If they are not removed, they will clog up the fusion cells.

The solar wind aids in purging the cells

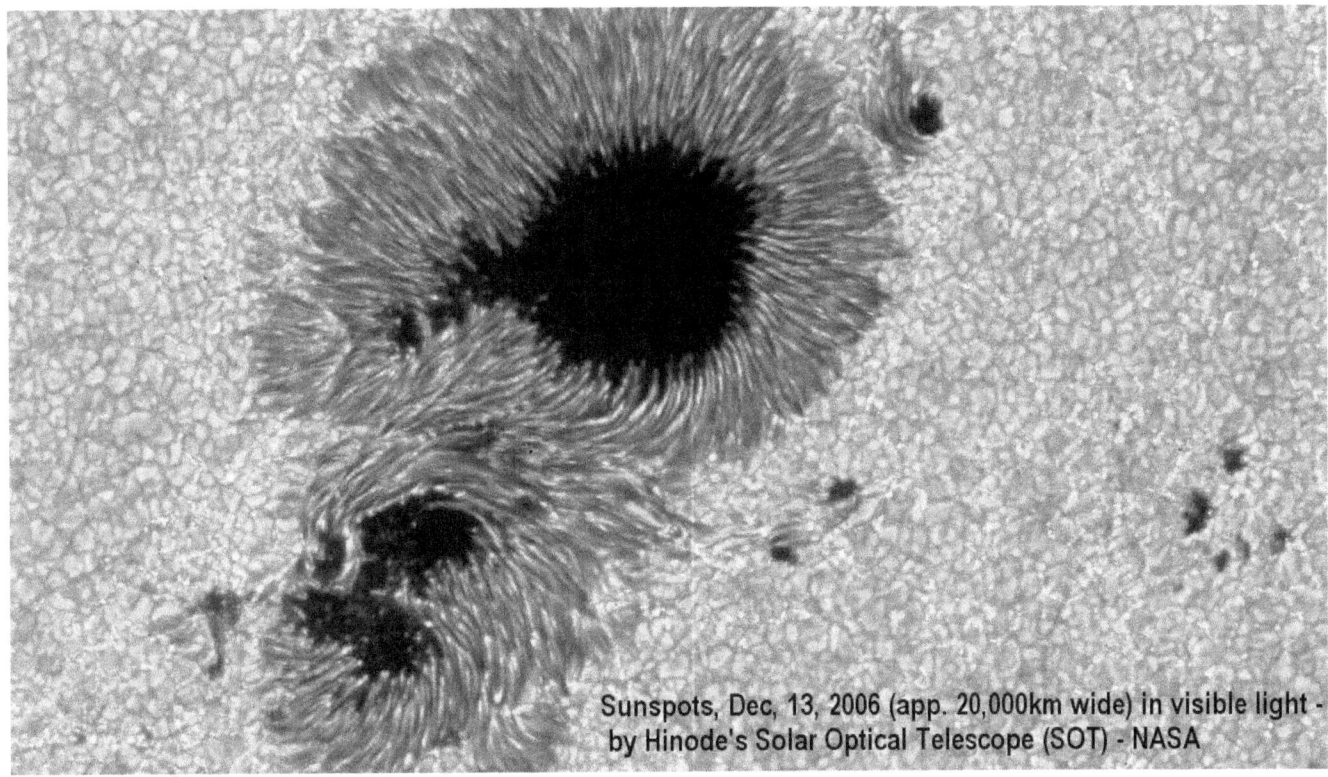

Sunspots, Dec, 13, 2006 (app. 20,000km wide) in visible light - by Hinode's Solar Optical Telescope (SOT) - NASA

The solar wind aids in purging the cells. When this service stops, the result is totally unpredictable. It may be minute. It may also be large. The uncertainty goes up 'through the roof.'

The cooling of the Sun will begin

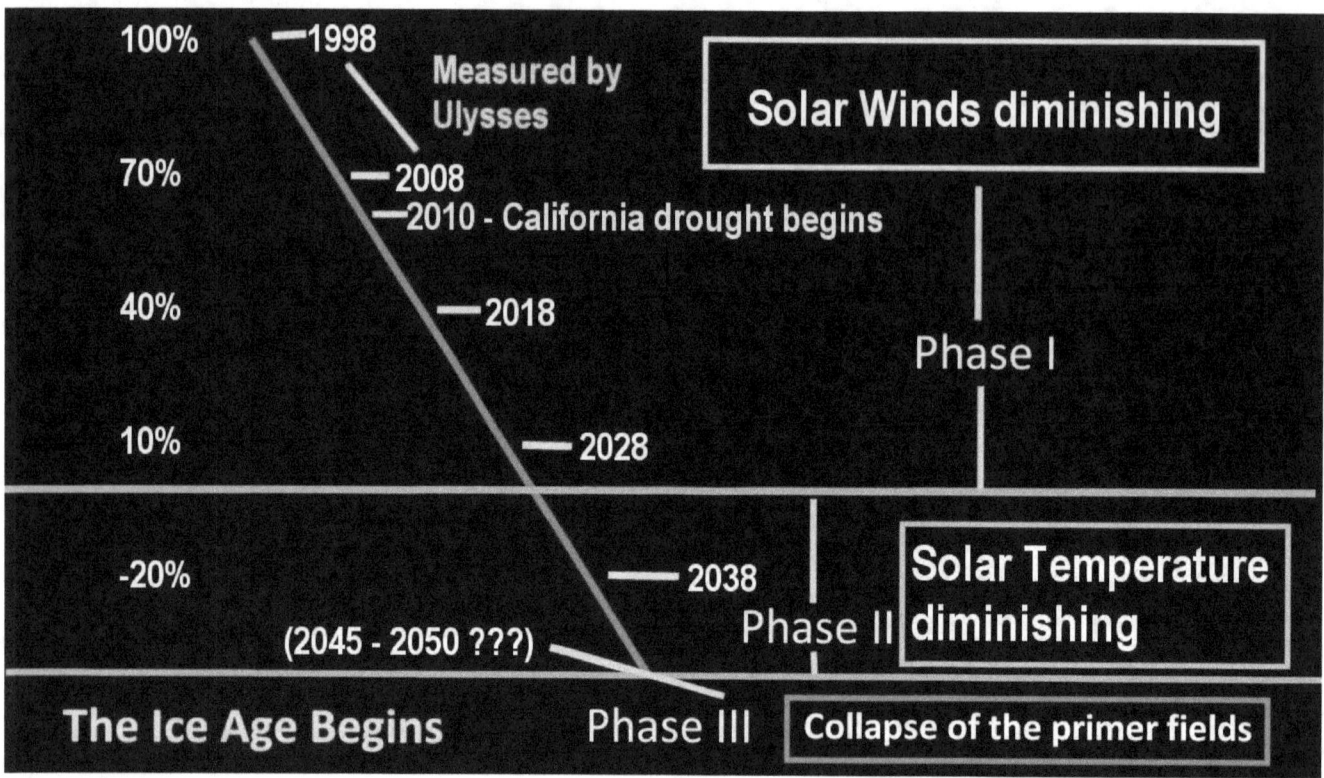

It is only predictable with certainty that the cooling of the Sun will begin when the solar wind stops, but not how rapidly it will progress, and how extensively it will affect all the other processes.

Thanks to the moderating effect of the solar wind

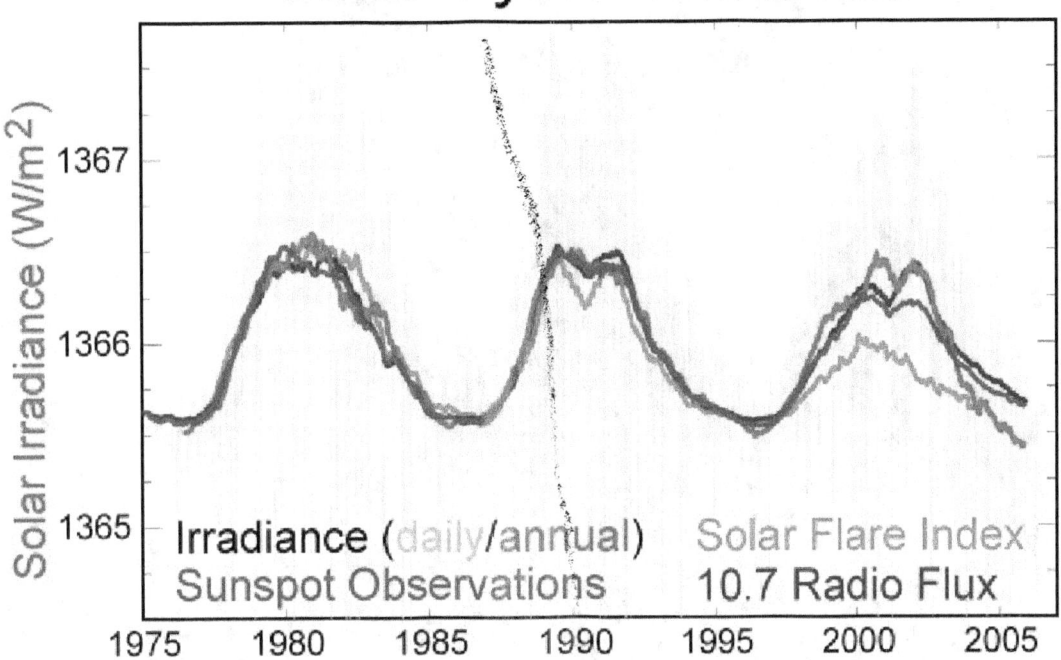

For as long as measurements have been made, the Sun's surface temperature didn't diminish significantly, thanks to the moderating effect of the solar wind. But when this moderating effect ends, we will see it reflected in the Sun becoming colder, and in reduced fusion actions becoming reflected in reduced plasma consumption, which reduces the Sun's sink effect that affects the flow-rate through the Primer Fields.

In a self-escalating rate of diminishment

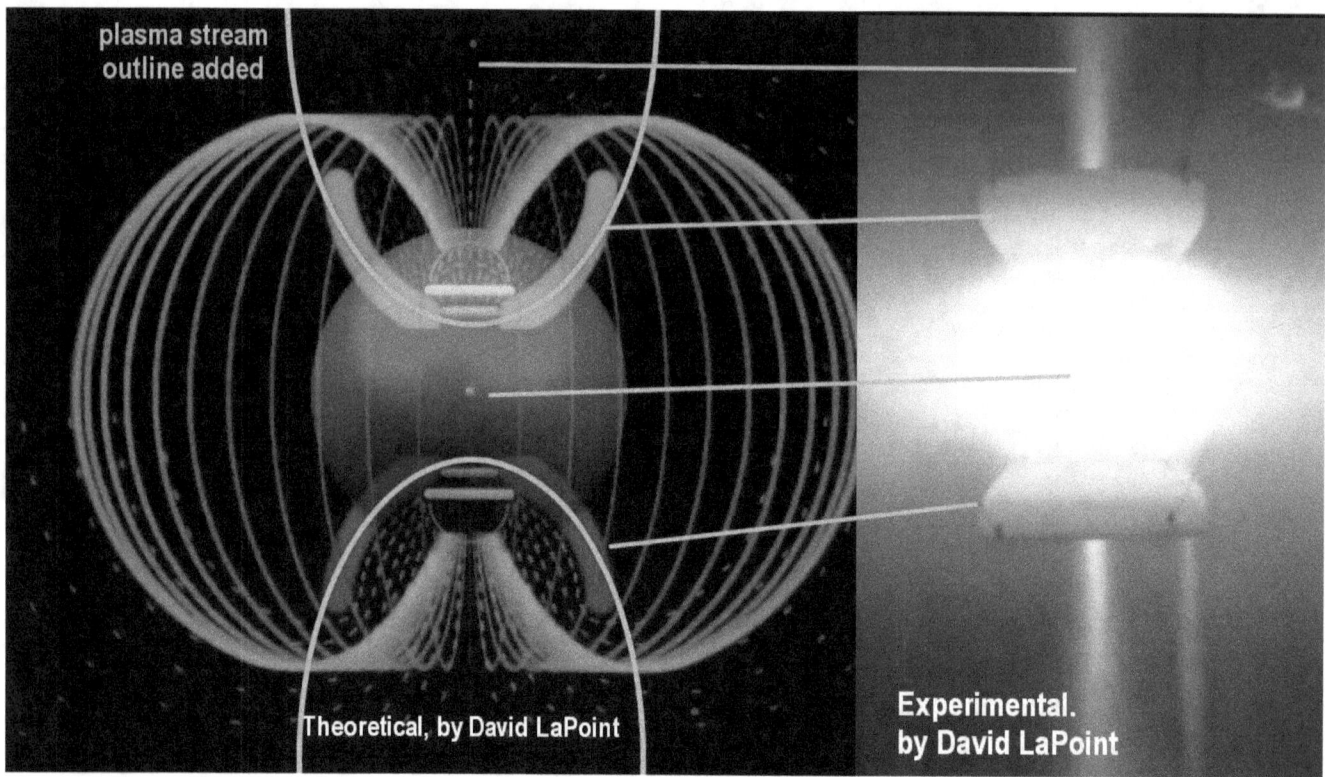

Then the reduced flow-rate, in turn, reduces the plasma concentration by the Primer Fields, which reduces the solar activity even more in a self-escalating rate of diminishment.

The self-escalating dynamic collapse

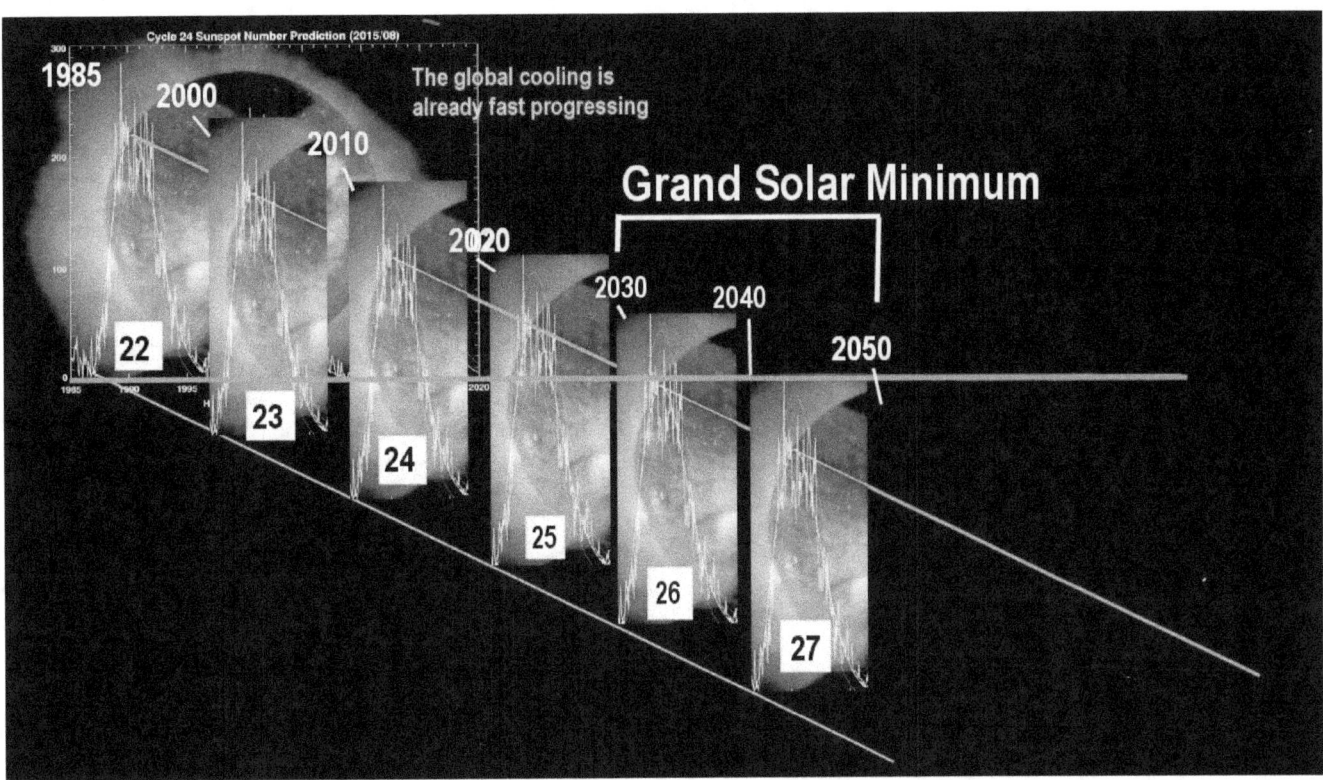

The self-escalating dynamic collapse furnishes the background for the coming Grand Solar Minimum, to become the next Ice Age.

For how long the Sun can grow colder

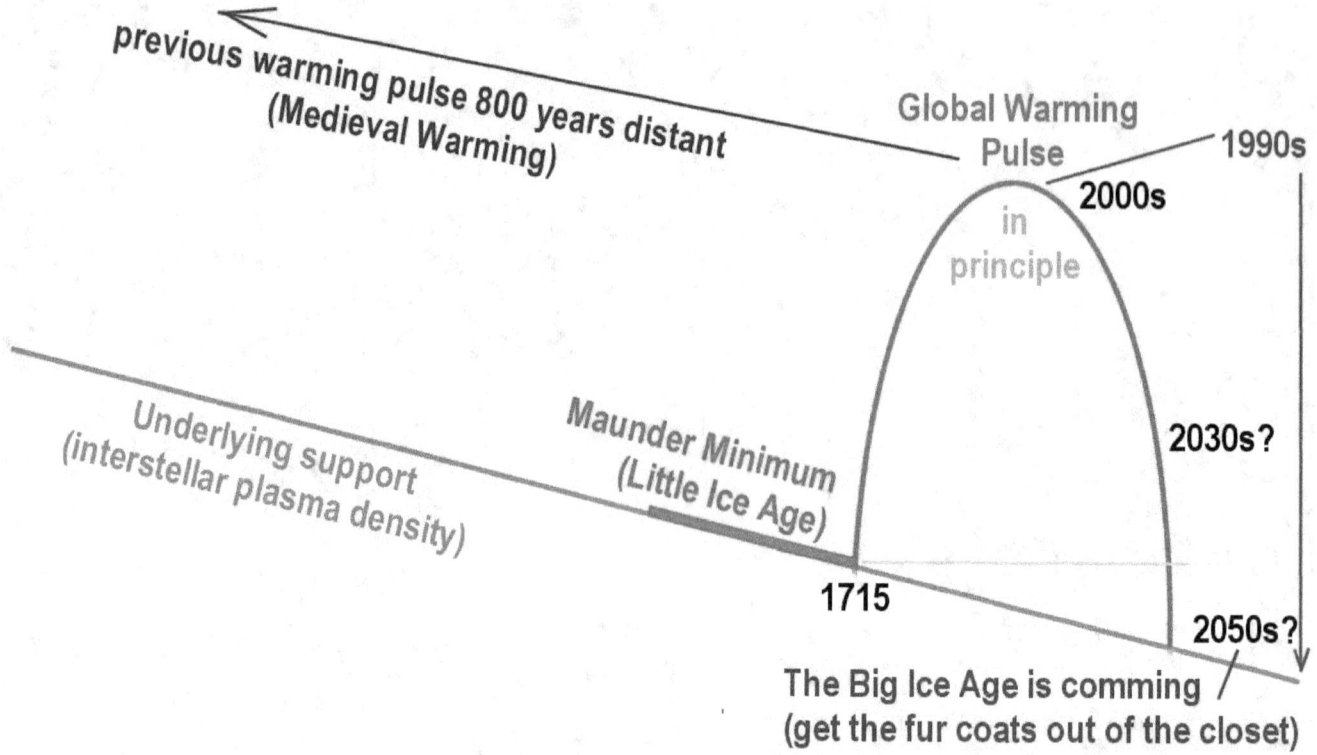

For how long the Sun can grow colder before the Primer Fields themselves collapse is wide open to uncertainty. The collapse could happen fast. It could happen during cycle 26, or it might linger on into the 2050s.

The closer we come to the big Ice Age phase shift

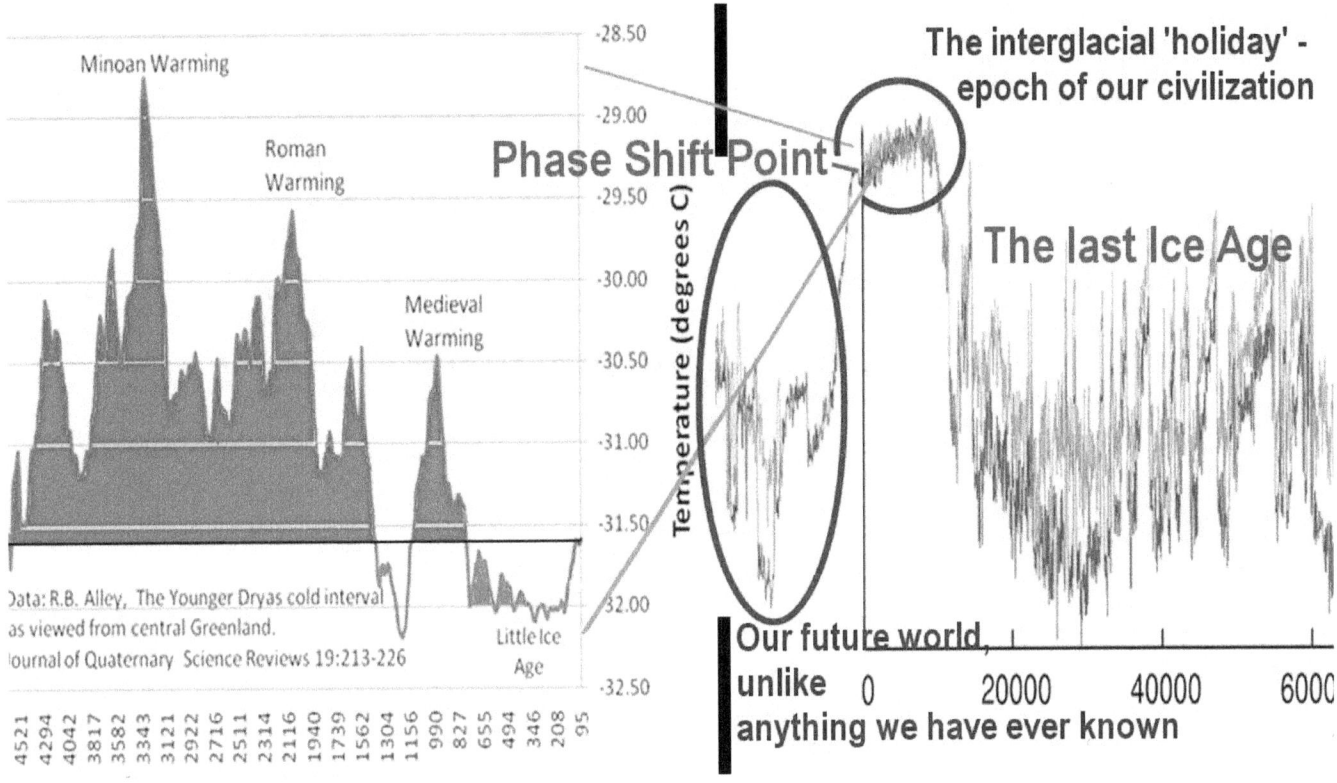

It seems that the closer we come to the big Ice Age phase shift point, the greater the uncertainties become that lead up to it. This is what we face in the near future. It may be nearer than we are willing to acknowledge.

➢ **The song of the warning bells**

> The song of the warning bells

The song of the warning bells

The consequences of the weakening Sun

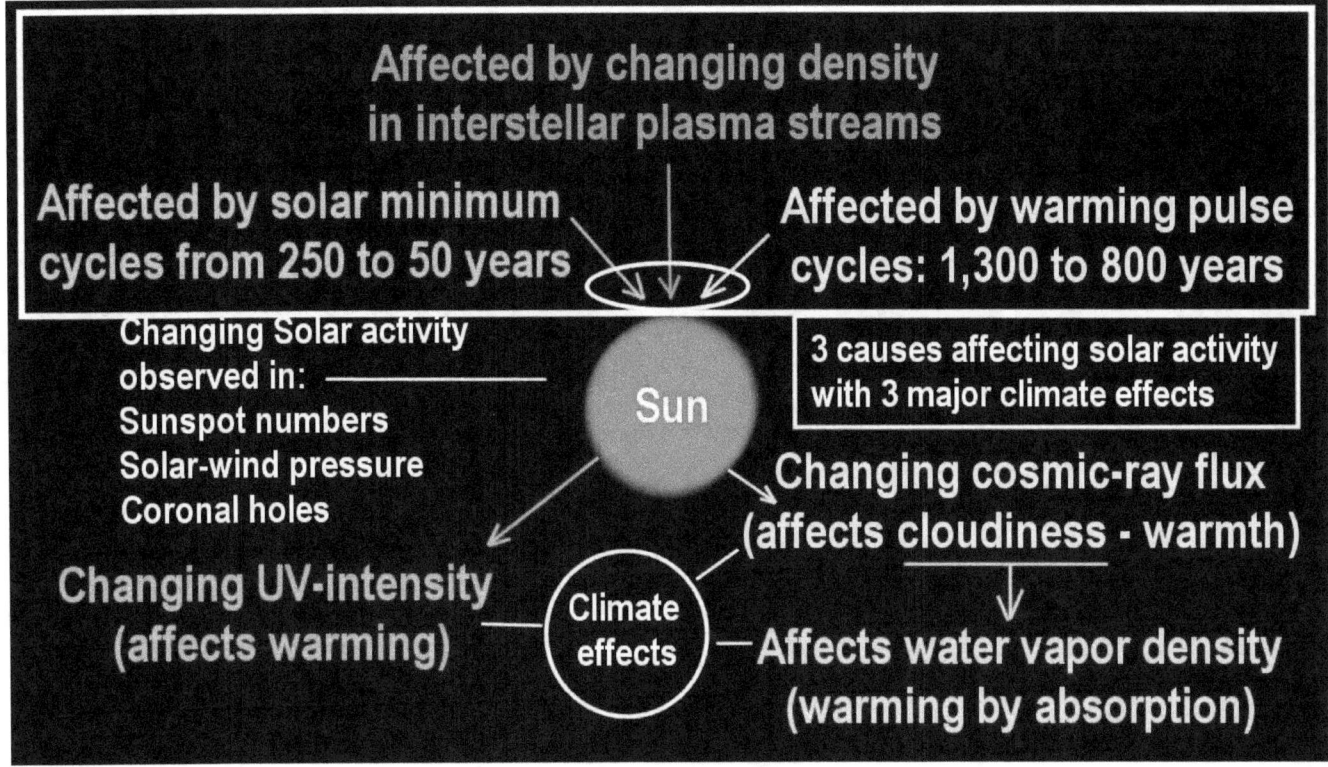

One thing is certain, that the consequences of the weakening Sun will become more-enormous the closer we come to the phase shift. The grand diminishing process has already begun with evermore devastating fringe effects along the way.

All the dynamic effects that we see combined, affecting the solar activity intensity, already have dramatic climate consequences that are in many ways more visible and hard than the changing Sunspot numbers are that we can still count,

As fringe effects, we can see by observation that changes in solar activity cause corresponding changes in UV-intensity in the sunlight, which affects the warmth of the climate on Earth.

Likewise the volume of solar cosmic-ray flux that we had measured historically, affects cloudiness in a big way, and thereby the warmth of the climate.

Changing cloudiness in turn affects the water vapor density in the air, which affects the Earth greenhouse warming by the heat absorption property of water vapor.

All of these effects become dramatically increased in the years and decades ahead as the solar system continues to weaken.

Sunlight includes a wide spectrum of light energy

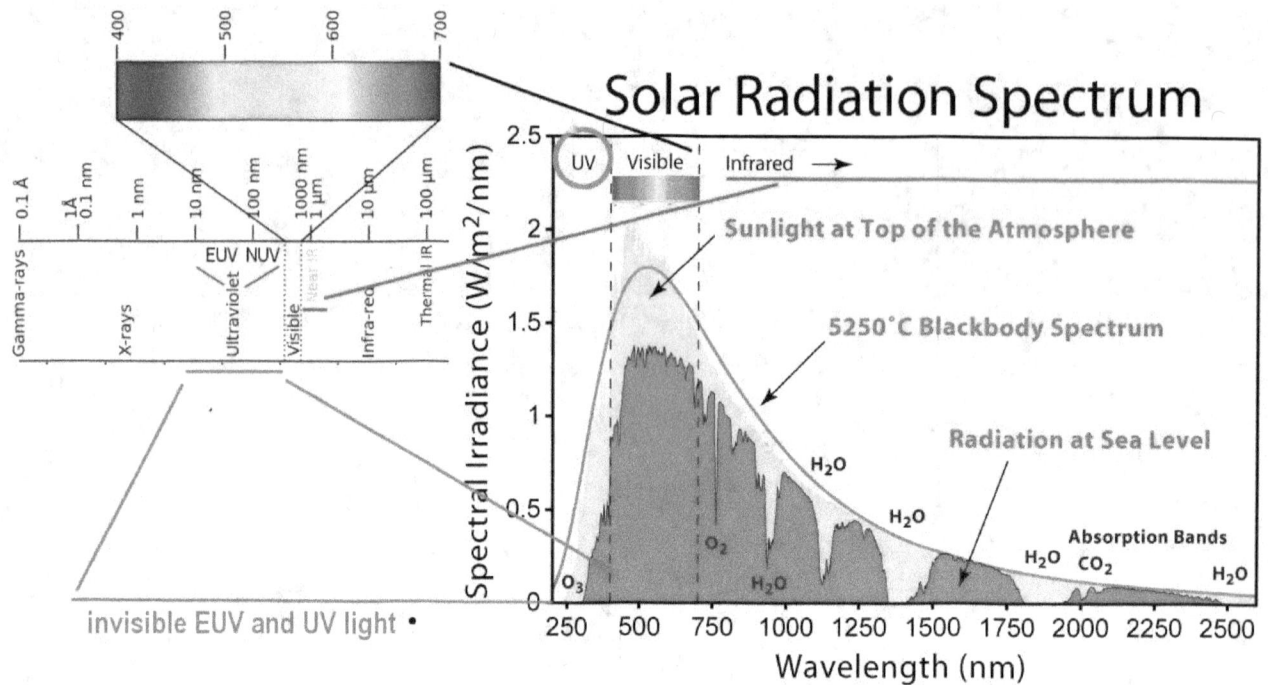

base image by "Global Warming Art images" - wikipedia

Sunlight includes a wide spectrum of light energy, of which only a small portion is visible. At the high-energy side of the spectrum, outside the visible range, above the violet, we find the ultraviolet 'color' that is invisible. This region is highly energetic. It causes sunburns and so on. And below the red, we find the infrared part of the spectrum, which we cannot see, but feel as heat. Each of these has a different effect on our climate, and the effects are getting bigger as fringe effects do in a dramatically changing energetic system.

Sunlight is absorbed in the atmosphere

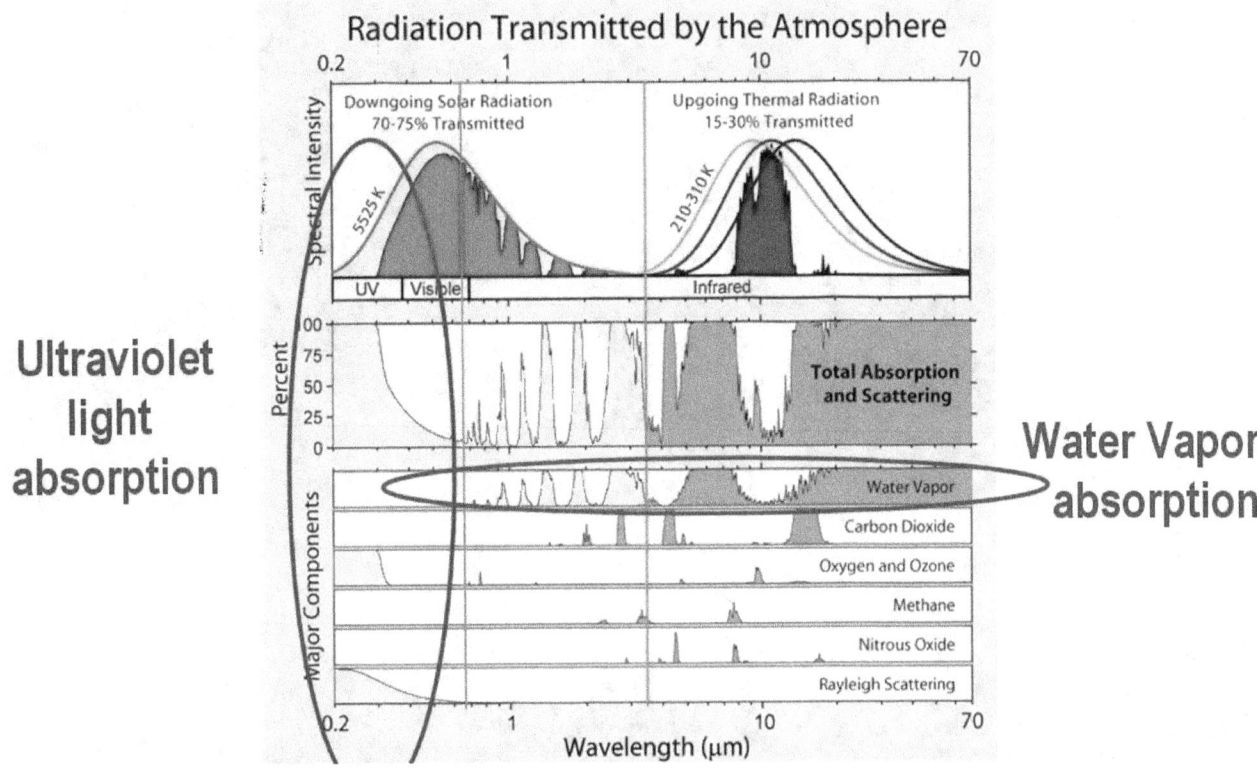

A large portion of the ultraviolet portion of the sunlight is absorbed in the atmosphere, typically by oxygen, and some UV energy is being scattered after being absorbed. The blue color of the sky results from the scattering effect that the UV radiation contributes to.

Both aspects, the direct absorption of UV light, and the scattering of it are significant contributors to the heat-budget of the atmosphere, and thereby to the greenhouse effect, and also to the heating of the soil.

In times when the sunspot numbers are high

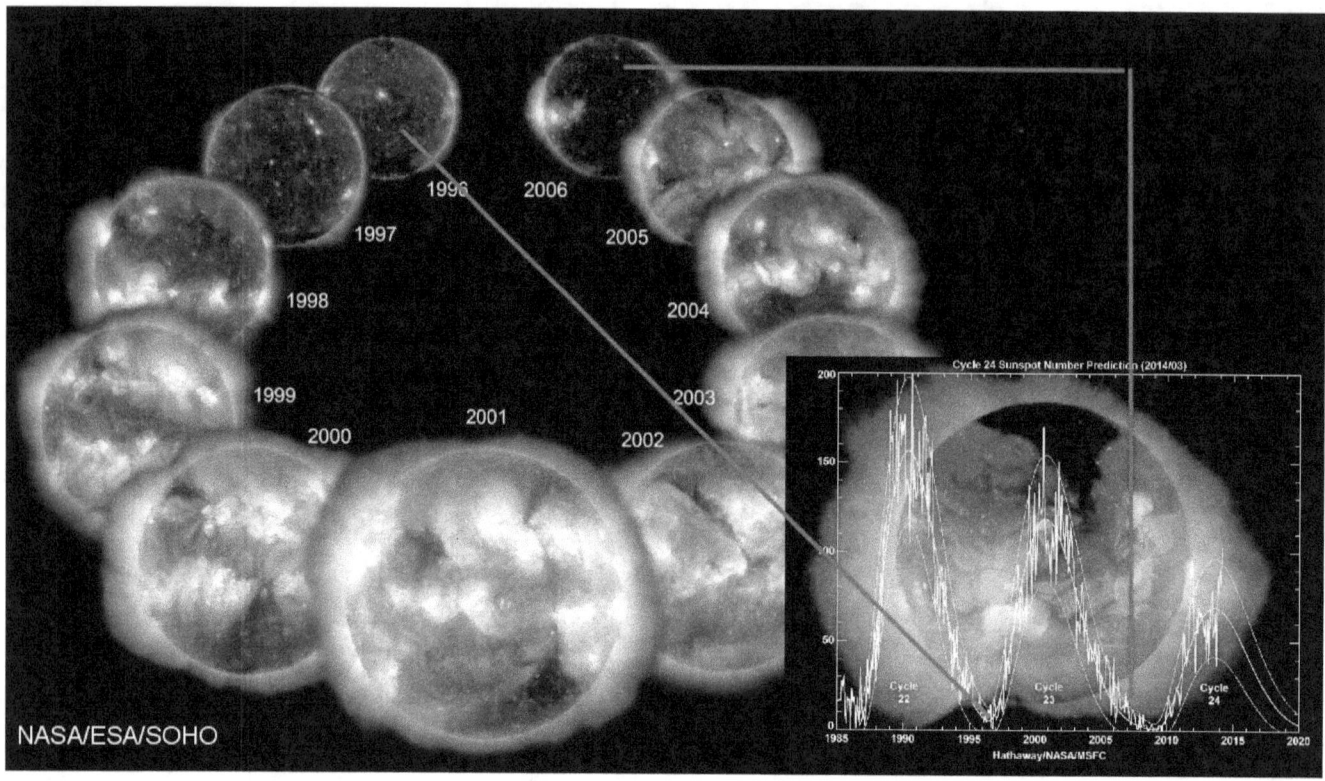

The UV light emission from the Sun varies dramatically, up to 10-fold, over the course of the solar cycles, as do the sunspot counts. While the sunspots have no effect on the climate directly, the changing level of solar activity does have a big effect, which the changing sunspot numbers are merely evidence of.

In times when the sunspot numbers are high, the Sun emits larger volumes of UV-light energy, and the Earth gets warmer. This means that changing solar activity is a strongly changing climate factor, affected by UV intensity fluctuations.

The infrared portion of the sunlight

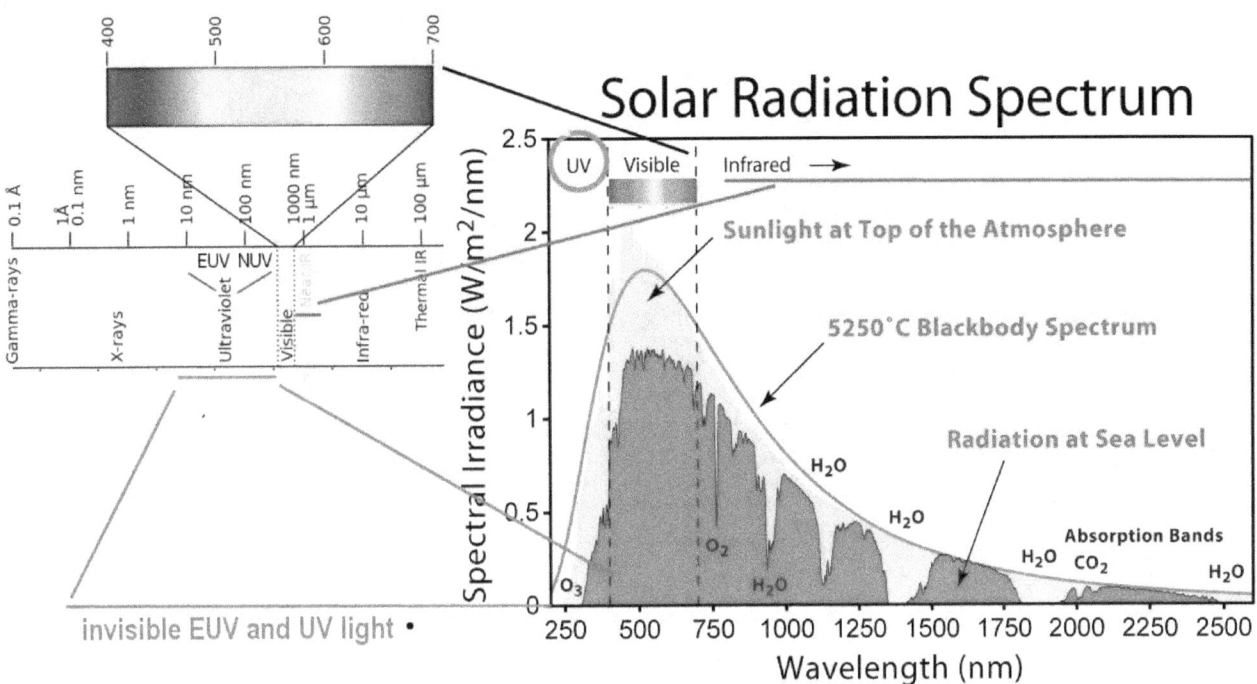

base image by "Global Warming Art images" - wikipedia

The infrared portion of the sunlight is likewise a climate factor. It contributes to the greenhouse effect with light being absorbed by water vapor in the atmosphere.

The water-vapor effect is the largest portion

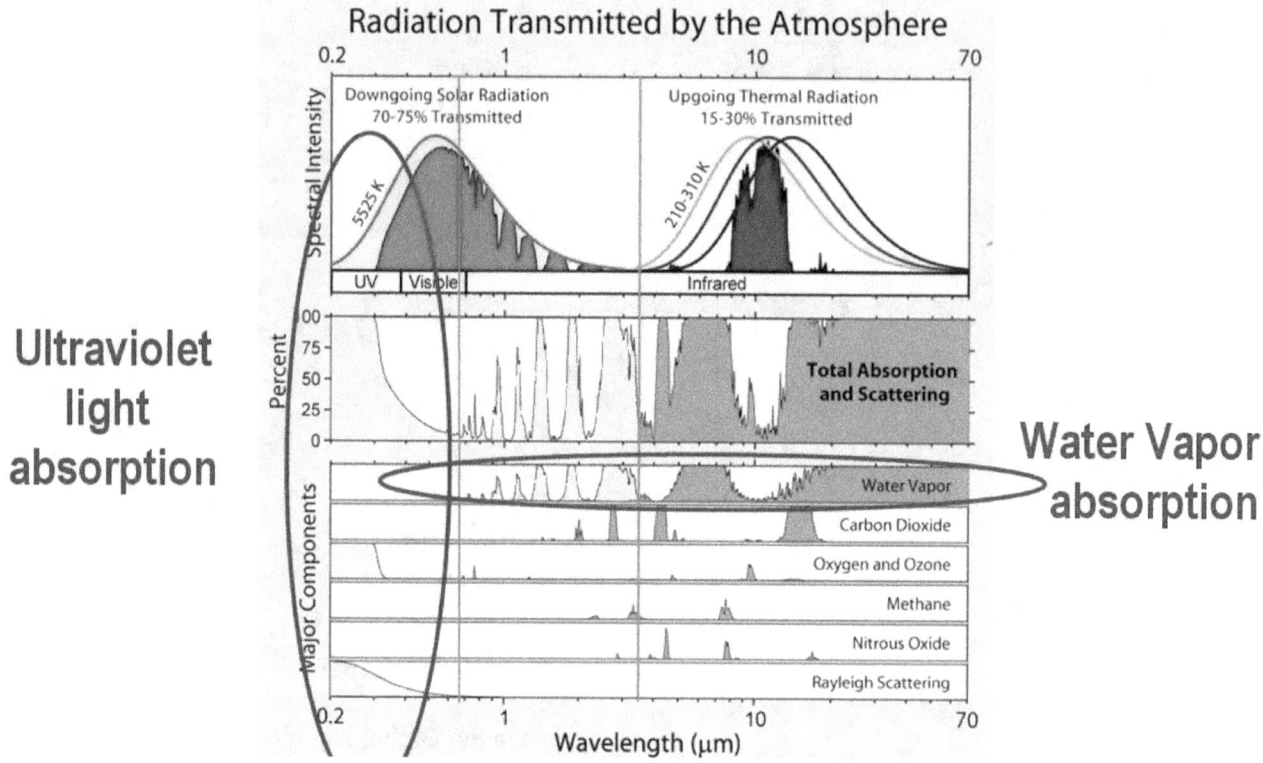

The water-vapor effect is said to be the largest portion of the solar radiation absorbed in the atmosphere. However, water vapor is not a constant factor. Its density changes with cloudiness. When clouds form, moisture is condensed out of the atmosphere, which reduces the greenhouse effect, subsequently. And cloudiness does change. It is highly effected by cosmic-ray intensity.

When solar activity is low,

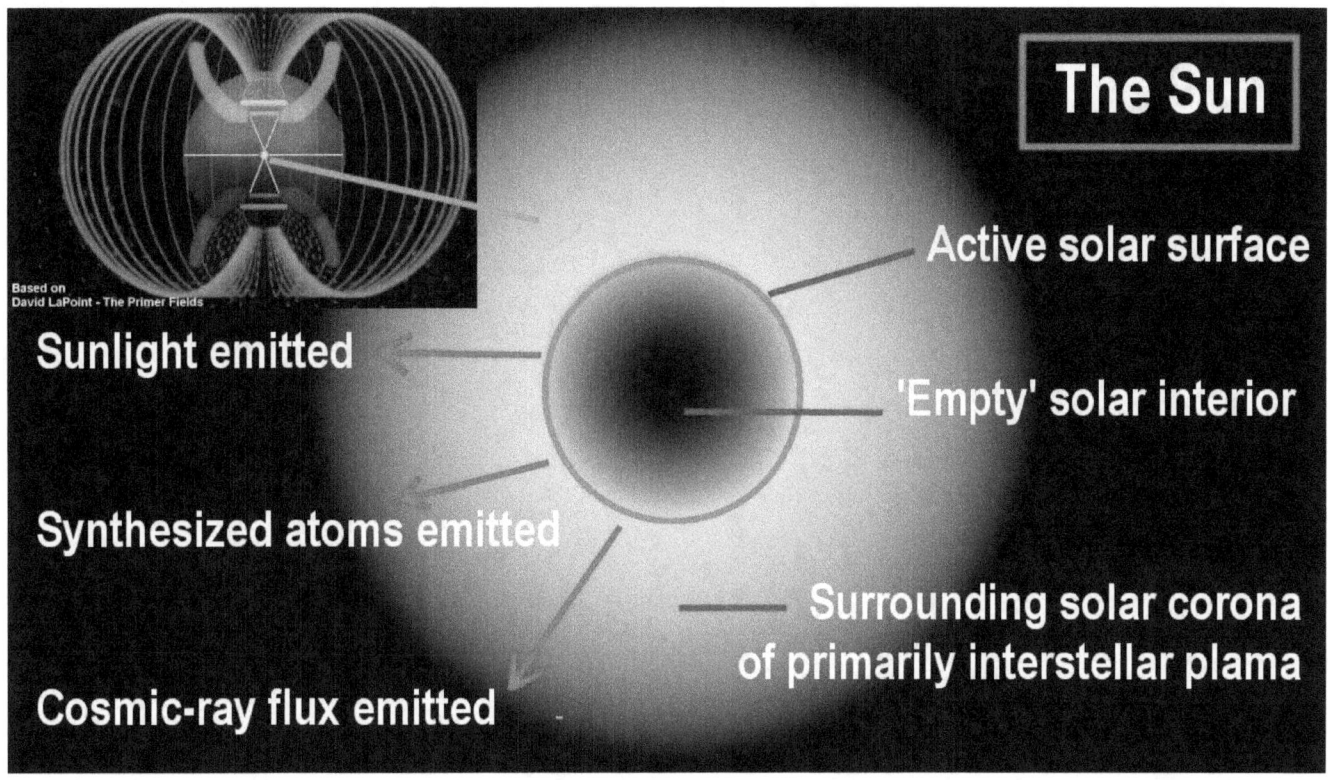

When solar activity is low, the Sun is surrounded by a weaker plasma corona. The weaker corona inhibits the Sun's cosmic-rays less. Consequently larger volumes are able to affect the Earth, which have an ionizing effect in the atmosphere.

Ionized elements in the air

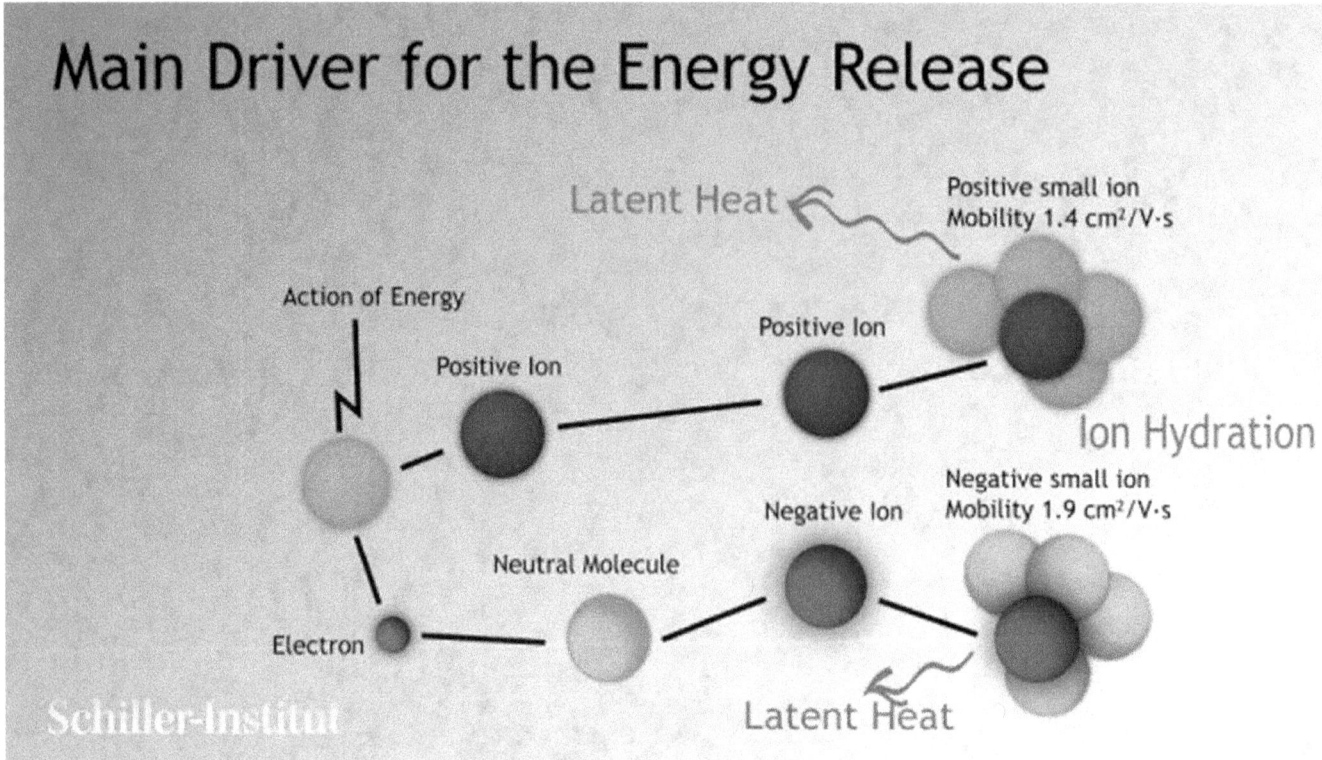

Ionized elements in the air are up to 100-times more attractive to water vapor. They increase cloud nucleation dramatically; increasing cloudiness.

Increased cloudiness, in turn

Increased cloudiness, in turn, reflects a larger portion of the incoming solar energy back into space, which thereby becomes lost to us, and the Earth becomes colder.

The effect of cosmic rays on cloud nucleation

CERN - CLOUD project - Jasper Kirkby

The effect of cosmic rays on cloud nucleation is often disputed, though it has been verified in principle with the CLOUD experiment at the CERN laboratory in Europe. A large test chamber had been filled with water-vapor. Then, artificial cosmic rays were injected.

When cosmic rays were injected

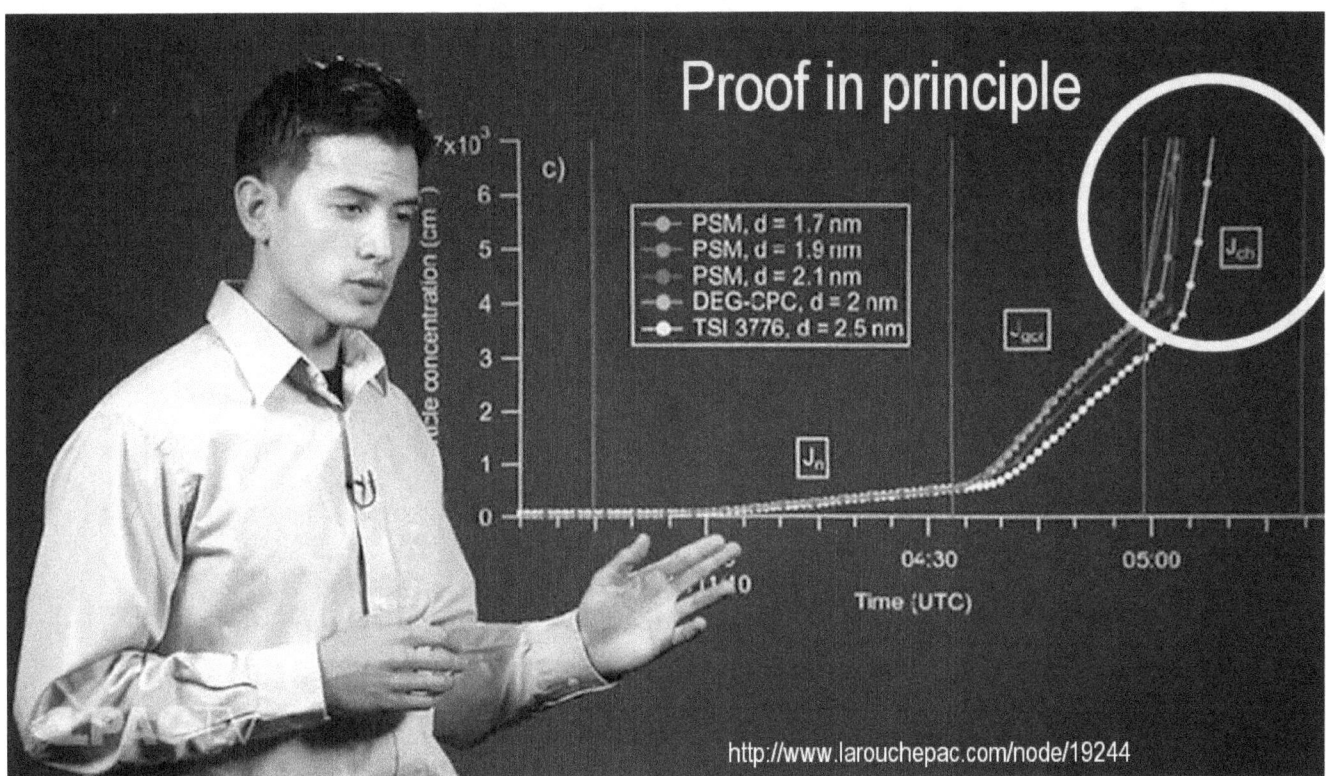

When cosmic rays were injected after the normal effects were measured, the rate of water nucleation went straight up and off the chart. The same happens when large volumes of cosmic rays are emitted from the Sun, which happens when the Sun is weak.

Inversely, when the Sun is strong, few cosmic rays are emitted, cloudiness is less extensive, and the climate is warmer.

Large climate fluctuations happen through the back door

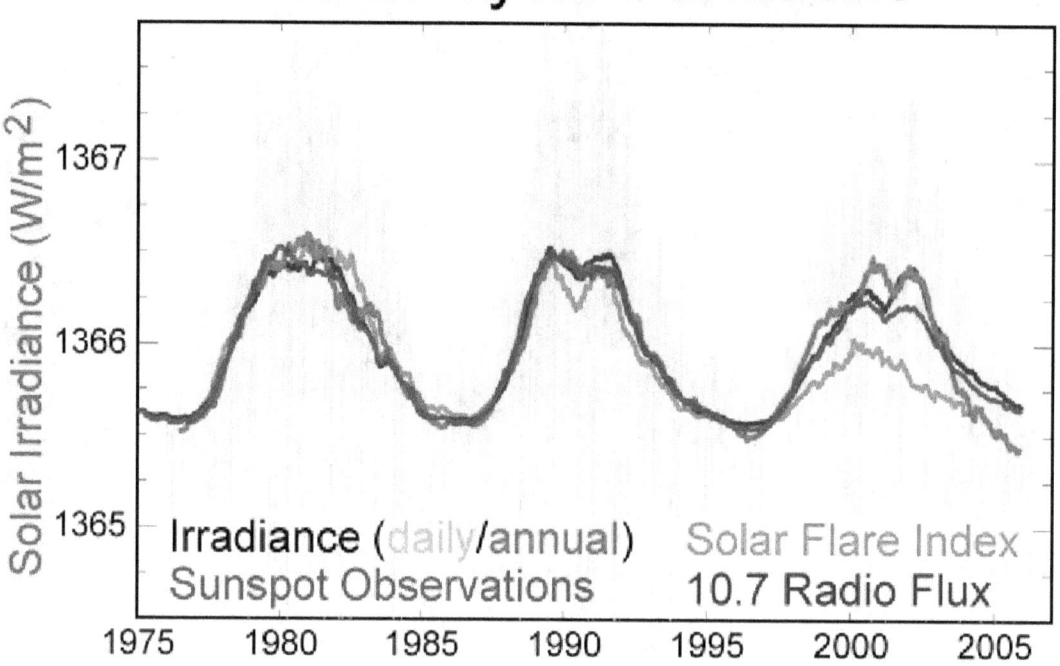

So it is, that by means of increased UV emissions from the Sun, and reduced cosmic rays for less cloudiness, the Earth becomes warmer.

The warming through the back door happens most-strongly when the sunspot numbers are high. This also means that the climate cools significantly when the solar activity is low.

All of these large climate fluctuations happen through the back door, while the front door, the Sun's surface temperature, still remains steady.

➢ **Not how we affect the climate**

> Not how we affect the climate, but how the climate affects us!

Not how we affect the climate, but how the climate affects us!

CO2 is comparable to a cat on the sidewalk

CO2 plays no role in the climate dynamics, because it is too feeble to affect anything, and is NOT affected by the Sun that is the climate master on the Earth. CO2 is simply a non-issue. It has no practical effect whatsoever. But the climate effects caused by the Sun do immensely affect us. That's the issue that counts.

CO2 is comparable to a cat on the sidewalk at the World-Trade towers in New York. The towers represent the UV and water-vapor greenhouse effects, and the cat represents CO2. Even if one would overfeed the cat so that it became a horse, it still wouldn't amount to anything in comparison.

A 10-times lower absorption coefficient

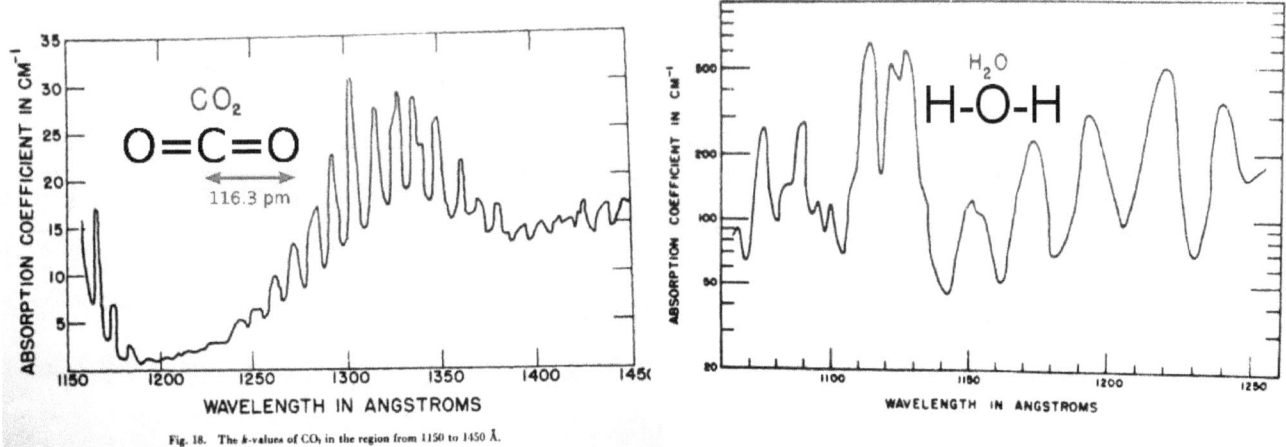

Compare the K values: 10 - 30 for CO2 vs. 100-600 for water vapor (H2O)

Fig. 18. The k-values of CO_2 in the region from 1150 to 1450 Å.

From a 1953 study by the Geophysics Research Directorate of the Air Force Cambridge Research Center Cambridge, Massachusetts - http://www.dtic.mil/cgi-bin/GetTRDoc?AD=AD0019700

CO2 has a 10-times lower absorption coefficient than water vapor.

CO2 is 100 times less dense than water vapor

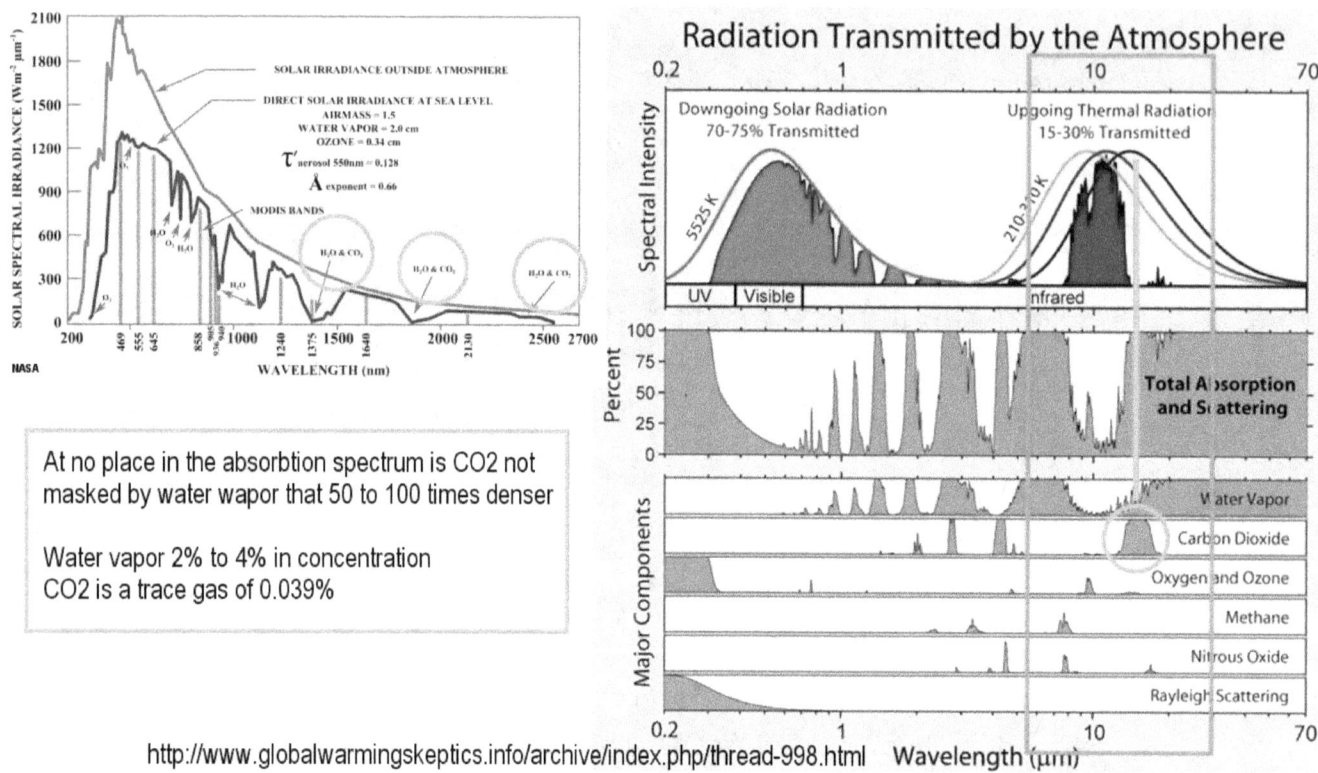

In addition, CO2 is 100 times less dense than water vapor, and responds only in a few narrow bands at the low-energy end of the spectrum, and even there, it is almost completely masked by water vapor, and is far overshadowed by the ever-variable factor of UV intensity and the light scattering effect. With all the factors added into the previous comparison, the 'cat' is most likely no bigger than a mouse, and this is so even without us considering the climate effect of cloudiness, which is also a big variable factor.

Greenhouse heat budget 45% of it as latent heat

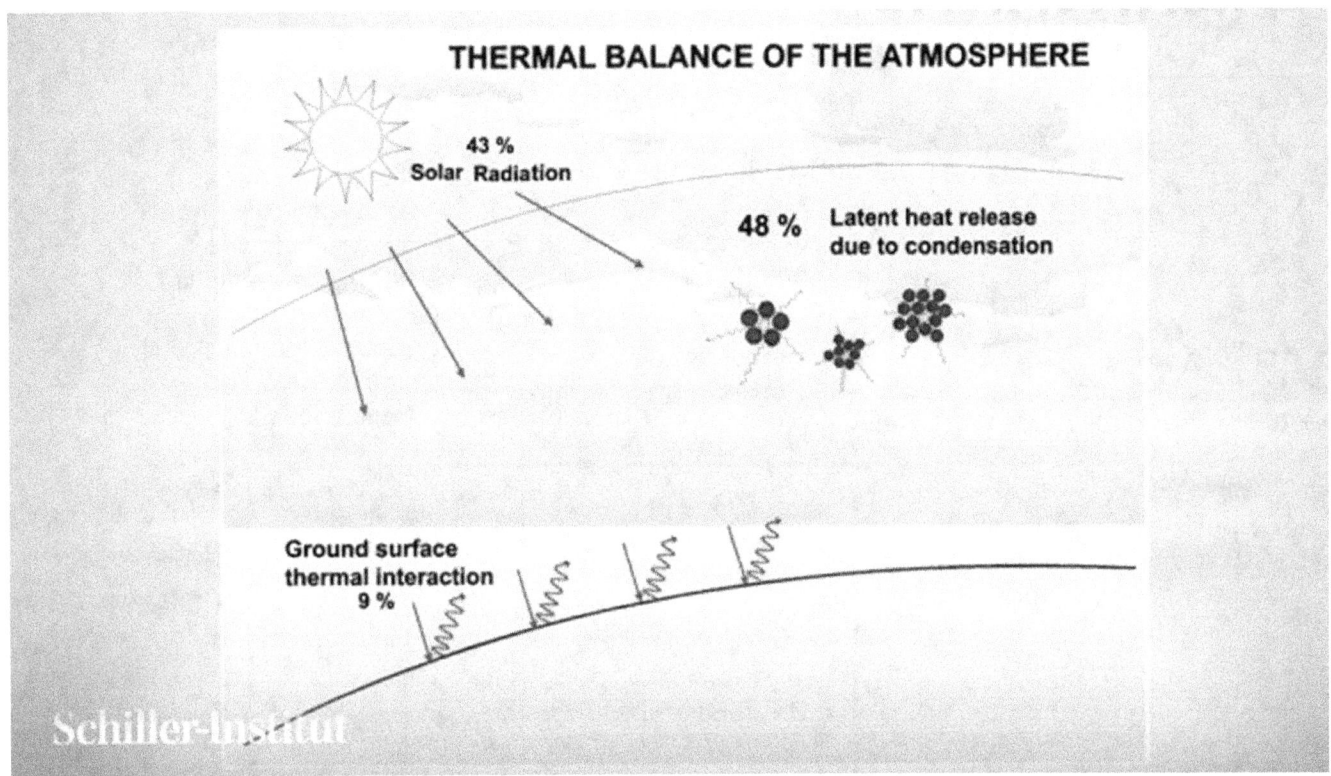

The greenhouse heat budget gets a large portion of its heat, up to 45% of it, from heat released by the process of cloud forming. When atmospheric water vapor condenses into clouds, the energy that had been absorbed to turn water into vapor, is becoming released as latent heat when the vapor condenses back into water. The cloud condensation, of course, reduces the water-vapor density, which reduces the water vapor's thermal absorption, which is a big factor of the greenhouse effect.

Since almost all of these factors are variable factors, we encounter a large number of uncertainties here, even while we feel their effects unmistakably.

Ominous 'writing on the wall'

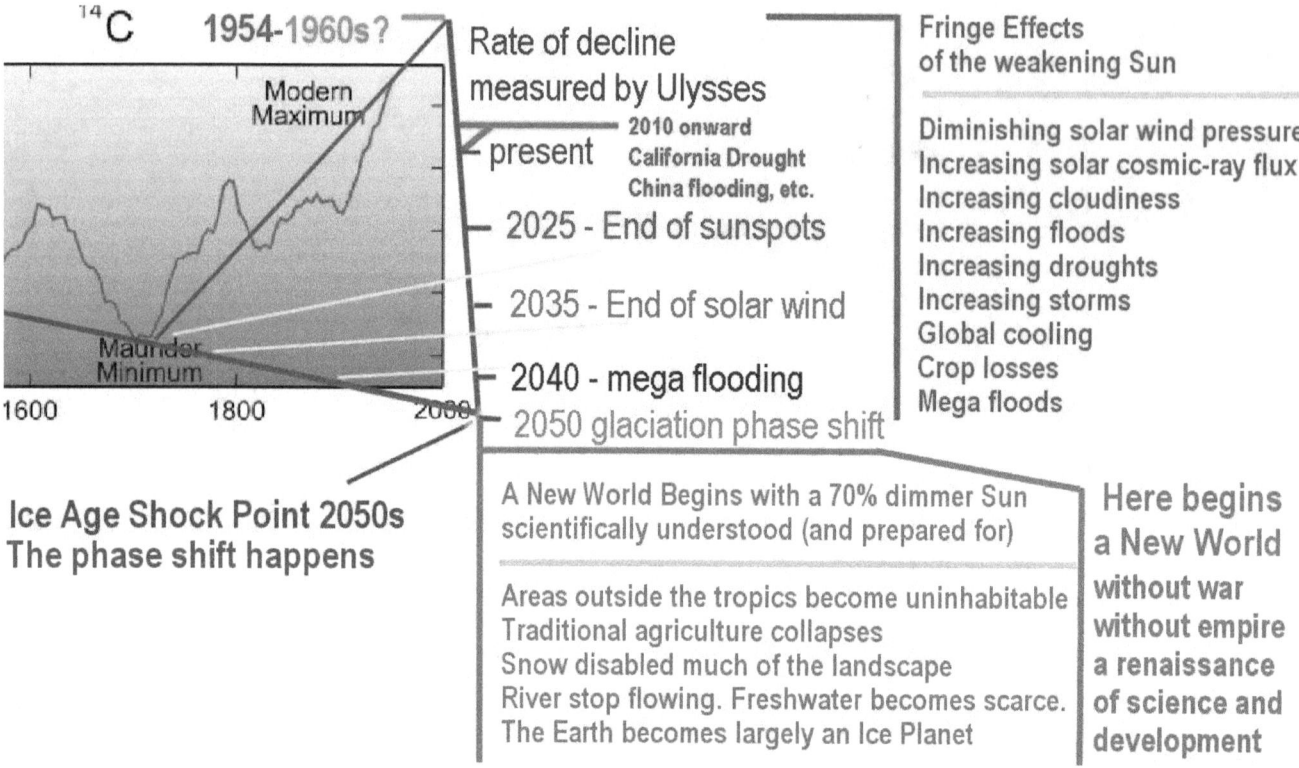

We see these increasing fringe effects as an ominous 'writing on the wall' in the form of increasing flooding, droughts, storms, snowfalls, general cooling, crop losses, forest fires, and in big examples such as the big California Drought, the big China Flooding events, worldwide mega-earthquakes and hurricanes, and so on. And we see all this increasing while the weakening of the solar system is still at the beginning stage. We have almost two decades yet to go on this worsening line till we get to the next Grand Solar Minimum that we won't likely recover from, after which the Ice Age begins and things get really bad.

When the world becomes largely uninhabitable, in part by the cold, and in part for the lack of freshwater, as the Ice Age unfolds, the term 'bad' is too mild.

➢ **To turn 'bad' into 'gold'**

> We have the option
> to turn 'bad' into 'gold'

We have the option to turn 'bad' into 'gold'

To build us a New World

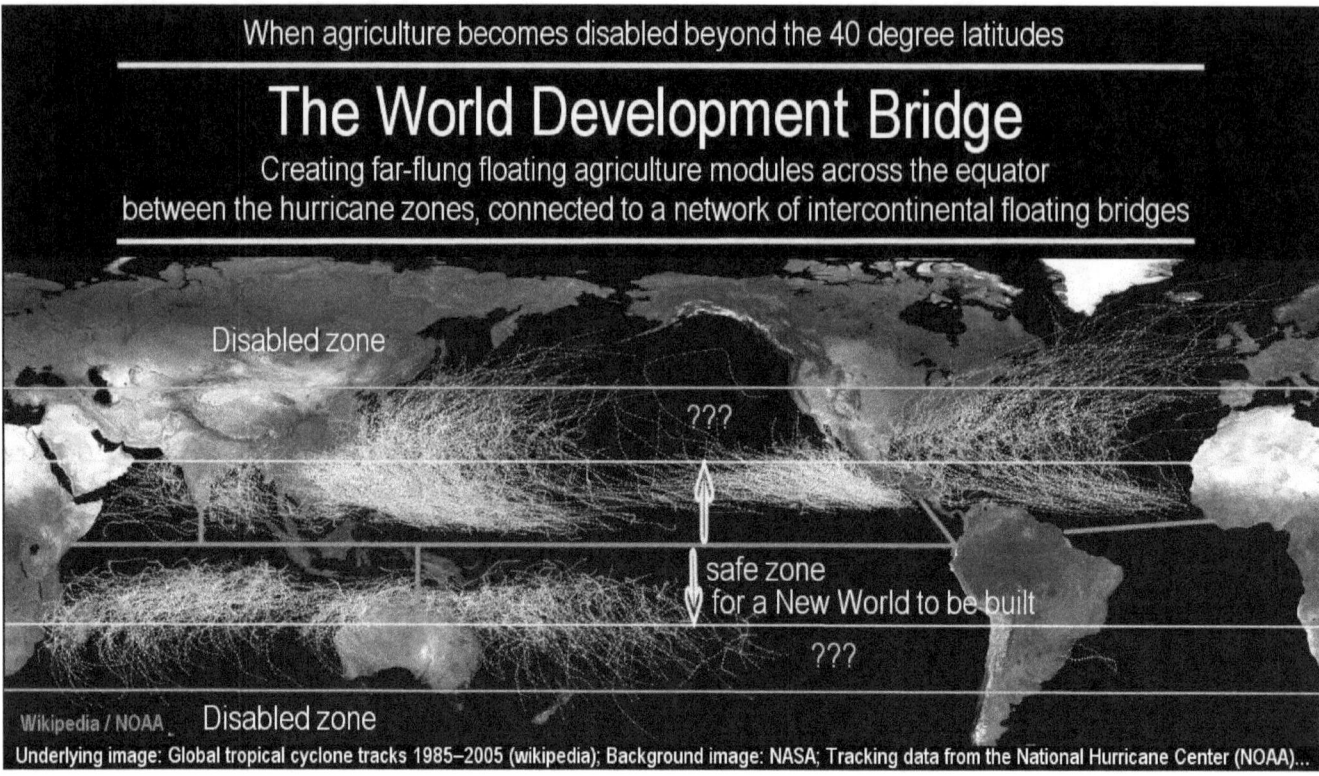

We have the option, as an intelligent humanity, to build us a New World that the expanding climate crisis cannot affect.

Life began in the sea

Life began in the sea. It may become our home again as a secure place in an uncertain future.

We can build secure agriculture afloat on the sea

We can build secure agriculture afloat on the sea, much of it indoors, serviced by a society with a new paradigm for living.

Living in thousands of new cities

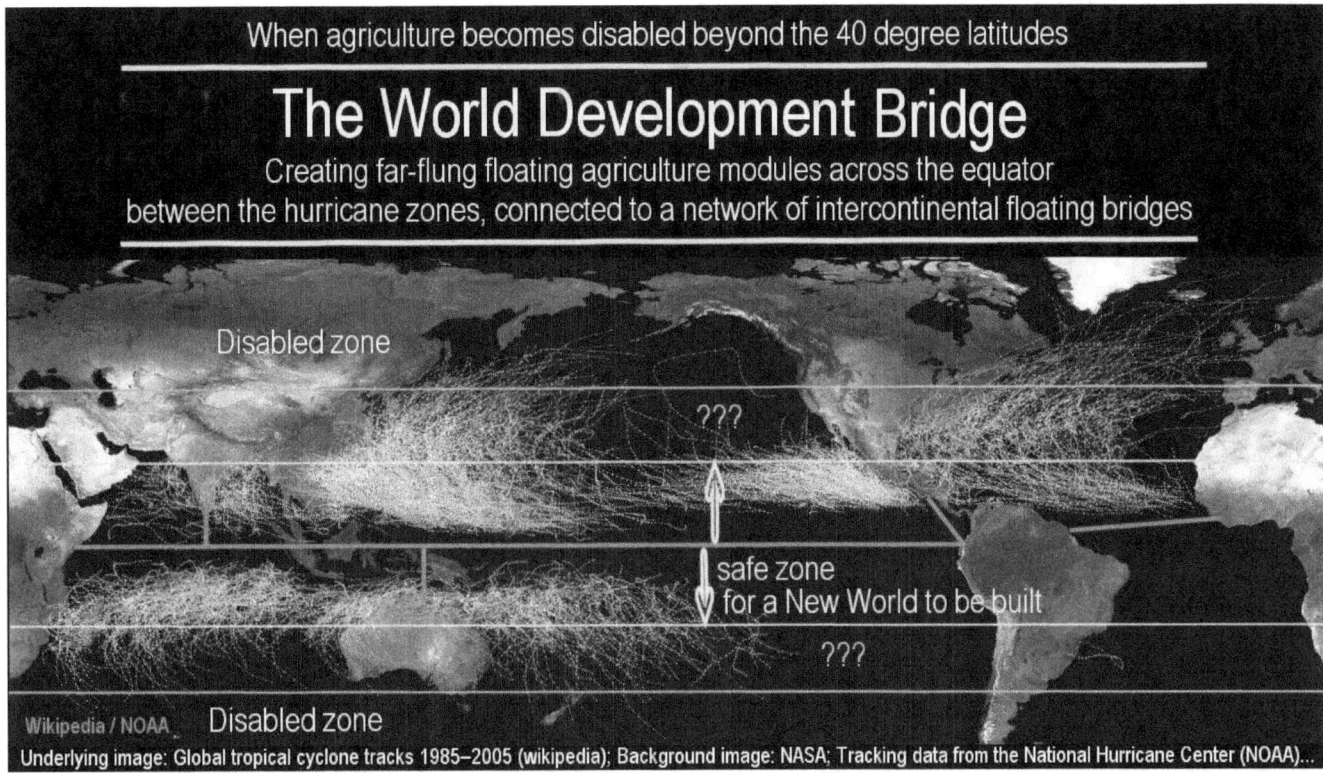

We will be living in thousands of new cities with secure and free housing, with efficient transportation infrastructures and energy infrastructures, in a world without war, without stealing, without want, and with a golden future to look forward to.

All nations working hand in hand

All nations working hand in hand

While this is still a dream, dreams can be made true with a little effort and commitment to it. Even while the world is already becoming dark, we can assure our children across the world that they will have a future, and we can prove our words by putting the spate into the ground.

We cannot escape the coming Ice Planet Earth

While we cannot escape the coming Ice Planet Earth, with absolute certainty, we can embrace it without fear, with the certainty in knowing that we, as human beings, are bigger - way bigger - than what the climate can bring.

Face the dimmer Sun with a song

We can face the dimmer Sun with a song and become our own light for the world.

We may not only walk on the water

We may not only walk on the water, but dance on it as we celebrate our self-made freedom, our freedom to live and love on the high level above the tumults of the small-minded conventionality where we are tied to the ghost of uncertainties.

When we raise ourselves up

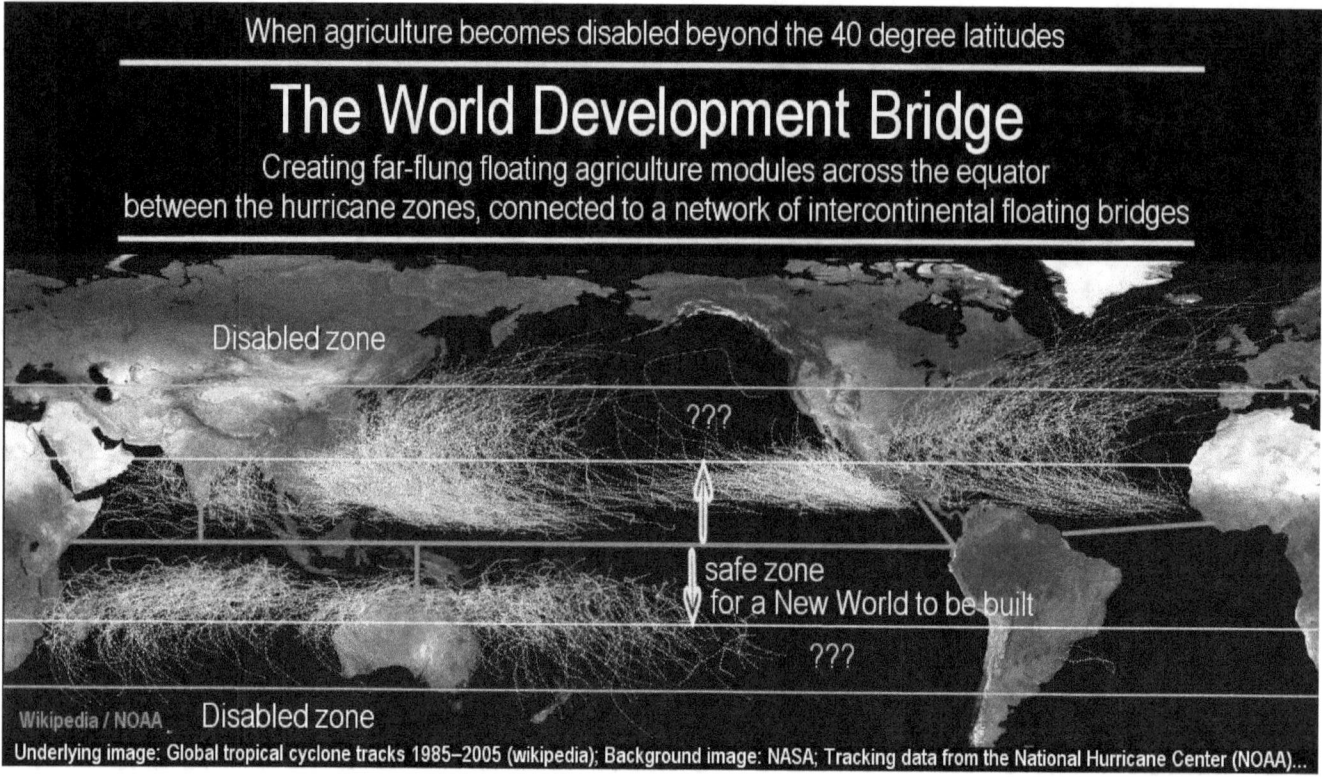

When we raise ourselves up and build that New World for us that we are fully capable of creating, then the long train of uncertainties ends. Then we will have a future.

The children of the world will be our joy and treasure

And then, without fail, the children of the world will be our joy and treasure.

The end

➢ More from the author:

14 Libraries of books and video productions

<u>Novels on Universal Love</u>, the greatest principle in civilization - **14 major novels**

Flight Without Limits (science fiction)

Brighter than the Sun (nuclear war avoidance?)

A series of twelve novels: **The Lodging for the Rose**
exploring the Principle of Universal Love

Book 1 - **Discovering Love**

Book 2 - **The Ice Age Challenge**

Book 3 - **Roses at Dawn in an Ice Age World**

Book 4 - **Winning Without Victory**

Book 5 - **Seascapes and Sand**

Book 6 - **The Flat Earth Society**

Book 7 - **Glass Barriers**

Book 8 - **Coffee Sex and Biscuits**

Book 9 - **Endless Horizons**

Book 10 - **Angels of Sex in Queensland**

Book 11 - **Sword of Aquarius**

Book 12 - **Lu Mountain**

<u>The Sex and Sacrament Project</u> - exploration stories from my novels - **11 books**

The Son of God

Impotence and Power

Self-Love and the Golden Hijab

Erica's Flower Garden

Helen a Healer

Brilliance of a Night

Gem of the Universe

The Sound of a Bird Woke Me

Between Ice and Spirit

Anton of Grace

Goodness of Living

The Kaleidoscope Project - mixed media of stories from my novels
- videos, PDF, audio

Discovering Infinity - developing history - 13 major research books:
A Research Book Series focused on scientific and spiritual development

Volume ii (Introduction) **Roots in Universal History** (Focus on Reality)

Volume 1A **The Disintegration of the World's Financial System** (Focus on Truth)

Volume 1B **Crimes Against Humanity** (Life Denied)

Volume 2A **Science and Christian Healing** (History as Truth)

Volume 2B **The Lord of the Rings' Metaphors**

Volume 3A **Universal Divine Science: Spiritual Pedagogical** (Structure for Discovery and Scientific Development - The Scientific Process to Know the Truth)

Volume 3B **Science and Health with Key to the Scriptures in Divine Science**

Volume 3C **Bible Lessons in Divine Science - 1898**

Volume 3D **Living in the Sublime**

Volume 4 **Light Piercing the Heart of Darkness** (The Demands of Truth and Justice)

Volume 5 **Scientific Government and Self-Government** (Platform for Freedom)

Volume 6A **The Infinite Nature of Man** (The Fourth Dimension of Spirit)

Volume 6B **Leadership** (The Spiritual Dimension of Leadership)

Cool Science of Kids - Illustrated Science - **interactive, videos, and 20 books**

War, Economics, and Nuclear War - scientific exploration - **10 videos**

Civilization - series focused on humanity - **10 videos**

Global Warming Doctrine - science videos - **12 videos**

Freshwater and Energy - science videos - **7 videos**

Christian Science explorations - **16 videos**

Books by Mary Baker Eddy - Christian Science - **16 on-line books**

Books by Rolf Witzsche on Christian Science - **9 Books**

The Giant PDF Library all transcripts of videos in PDF form

For links, please see: http://www.ice-age-ahead-iaa.ca

The projects are designed to draw the riches of our humanity into the foreground **towards a New Renaissance**, in order that their light may out-shine the systems of empire that are erroneously accepted, including the follies of war, terror, looting, economic destruction, science-perversion, and policies for depopulation.
Rolf A.F. Witzsche